Design of Demining Machines

Dinko Mikulic

Design of Demining Machines

Dinko Mikulic
Velika Gorica
Zagreb
Croatia

ISBN 978-1-4471-4503-5 ISBN 978-1-4471-4504-2 (eBook)
DOI 10.1007/978-1-4471-4504-2
Springer London Heidelberg New York Dordrecht

Library of Congress Control Number: 2012945107

Printed on acid-free paper

Springer is part of Springer Science+Business Media (www.springer.com)

Foreword

In constant effort to eliminate mine danger, the international mine action community has developed the safety, efficiency, and cost-effectiveness of clearance methods. Demining machines have become necessary when conducting humanitarian demining. In humanitarian demining practice, mechanization of demining makes a difference, providing higher safety and greater productivity.

This book is based on the real identity of demining machine development, which proved successful in the implementation of mechanic technology and achieved results of clearing mine suspected areas. It presents the methodology of design and the demining machine testing. Based on that, the demining machines produced and used in the world for the purpose of eliminating mine threat have been developed. This is the result of understanding their role in conditions of humanitarian demining.

We would like to thank the author for making this unique publication. It should help demining machine development in the world and their wider use in humanitarian demining.

Vladimir Koroman
Nikola Gambiroza

Preface

Design of Demining Machines describes the development and testing of modern demining machines in humanitarian demining. Machines and other equipment are mechanizing the demining process, thus replacing human involvement in this dangerous activity, and therefore reducing risk. Modern humanitarian demining cannot be imagined without the use of machines because this reduces time and costs for demining.

Relevant data for design of demining machines have been provided. Out of different humanitarian demining machinery, this book provides description of the perspective machinery that draws attention as well as some innovative and inspiring development solutions. It also describes the development technologies, companies, and projects. The provided data is gained through genuine research and development. Some data were excerpted from scientific research projects, scientific papers, and producers' material. Emphasized were also drafts of standards for testing and evaluation of demining machines. This book is primarily tailored as a text for the study of the fundamentals and engineering techniques involved in the calculation and design of demining machines. Much attention has been given to the dynamic processes occurring in machine assemblies and components, as well as in demining machine as a whole. This enables full estimate of the effects of various design factors and proper selection of optimal parameters for designing the demining machine.

I hope this book will contribute to the spreading of knowledge for safer and faster removal of mine threats in humanitarian demining.

Zagreb, May 2012 Dinko Mikulic

Acknowledgments

I would like to use this opportunity to express my appreciation to all my associates who have helped in the research of demining machines and equipment, especially to Mr. Ivan Steker for his help during the preparation of this book. I would also like to express my gratitude to Mr. Al Carruthers / GICHD for the guidance in preparing the book for publication.

Many thanks go to Mr. Vjekoslav Majetic and Mr. Tihomir Mendek from the DOK-ING Company for their assistance during the preparation of this book. I would also like to express my gratitude to the reviewers Vladimir Koroman, Ph.D. and Nikola Gambiroza, Ph.D.

Let me also thank CROMAC, CROMAC-CTDT, and DOK-ING Company in particular, as well as the Croatian Standards Institute, for providing their permission to use their own specific material. Finally, I would like to thank all the other demining companies and associates for their valuable suggestions.

Contents

Chapter 1
Humanitarian Demining Techniques

1.1 Landmines Threats

An estimated 100 million landmines are embedded across our planet [1]. Landmine monitor has identified at least 84 countries and eight areas contaminated with landmine and unexploded ordnance (UXO) in 2005. Of the affected countries, 54 are States Parties to Mine Ban Treaty.

Majority of these landmines are small anti-personnel mines. Landmines kill and wound over 20,000 people each year [2]. The 12 most heavily mine-affected countries in the world, according to a UN study, are: Afghanistan, Angola, Bosnia-Herzegovina, Cambodia, Croatia, Eritrea, Iraq (Kurdistan), Mozambique, Namibia, Somalia, Nicaragua and Sudan. These countries together account for almost 50 % of the landmines currently deployed in the world and also suffer the highest number of landmines causalities. As estimated 45–50 million landmines infest at least 10 million sq. km of land around the world.

In many mine-affected countries, however there is still a lack of knowledge as to the extent of the problem and detailed information as to the exact location of the mines. The total number of landmine survivors continues to grow as new casualties are recorded in every region of the world. Nationals from more countries are also killed or injured by landmines while outside their countries—as peace-keepers, tourists, or humanitarian and development workers.

1.1.1 Countries Affected to Various Degrees by Mines and Unexploded Ordnance

Europe/Central Asia

Albania, Armenia, Azerbaijan, Belarus, Bosnia and Herzegovina, Bulgaria, Croatia, Cyprus, Czech Republic, Denmark, Estonia, Georgia, Greece, Kyrgyzstan, Latvia, Lithuania, Moldova, Russia, Slovenia, Tajikistan, Turkey, Ukraine, Serbia

D. Mikulic, *Design of Demining Machines*,
DOI: 10.1007/978-1-4471-4504-2_1,

Africa

Angola, Burundi, Chad, Congo (Brazzaville), Democratic Republic of Congo, Djibouti, Eritrea, Ethiopia, Guinea-Bissau, Kenya, Liberia, Malawi, Mauritania, Mozambique, Namibia, Niger, Rwanda, Senegal, Sierra Leone, Somalia, Sudan, Swaziland, Tanzania, Uganda, Zambia, Zimbabwe

America

Chile, Colombia, Costa Rica, Cuba, Ecuador, Guatemala, Honduras, Nicaragua, Peru

Asia/Pacific

Afghanistan, Bangladesh, Cambodia, China, India, Korea, Democratic People's Republic of Korea, Republic of Laos, Mongolia, Myanmar (Burma), Nepal, Pakistan, Philippines, Sri Lanka, Thailand, Vietnam

Middle East/North Africa

Algeria, Egypt, Iran, Iraq, Israel, Jordan, Kuwait, Lebanon, Libya, Morocco, Oman, Syria, Tunisia, Yemen

«Hidden Killers» Statistics

- Effectiveness of demining is a constant problem, almost irresolvable comparing to number of mines set. According to some estimation, demining lasts hundred times longer than setting mines.
- Anti-personnel mines (AP) are usually hidden and ambushed with light activation stress ($\approx$10 daN) or by wire. Man with its weight cannot activate anti-tank mines (AT) ($\approx$300 daN), but AP and AT mines can be placed together which presents more danger.
- 100 million landmines lie in the ground in the world. It will cost $33 billion to remove only these mines (if no others are planted), which under present demining rates will take more 100 years to clear.
- There are 100 million landmines stockpiled around the world.
- 70 people are killed or injured every day by landmines 20,000 people per year.
- 300.000 children are severely disabled because of mines.
- Half of all adults who stand on a mine die before they reach hospital. Children, because of their size, are more likely to die from their injuries.
- Mines can cost as little as $3 to make and over $1,000 to clear.
- For every hour spent in sowing mines, over 100 h are spent de-mining.
- Demining is very dangerous. One accident occurs for every 1–2,000 mines removed.
- Victims need twice as many blood transfusions as people injured by bullets or fragments. Number of units of blood required is 2 to 6 times greater than that needed for other war casualties.

- Each prosthetic cost around US $3,000 per amputee in developing countries. For the 250,000 amputees registered worldwide this means a total expenditure of $750 million.
- The relevant measure of the mine problem is the number of square kilometers of land cleared of land mines, without human casualties.

Landmine Monitor Reports, UN and GICHD talk about AP mines as "a wide and long term pollution of environment and high level of mortality". AP mines do not have integrated self-destruction device, and jeopardize lives of millions of people each day. These are basic reasons to prohibit the production and use of AP mines. In Ottawa, in 1997, an international convention on prohibition of use of AP mines was inaugurated (122 countries), titled ***Convention on the Prohibition of the Use, Stockpiling, Production and Transfer of Anti-Personnel Mines and on their Destruction***. With this Convention, AP mines are placed on a list of weapons prohibited for use in armed conflicts, and in International Law of War mines gained the same status as other prohibited weapons of mass destruction, such as biological and chemical weapons (war gases).

1.1.2 LandMine Obstacles

From military point of view, land mines can be AP and AT mines. Here is provided a basic military use of these mines, in order to be familiar with their planting and possible ways of removing them. Also, a reference of their military uselessness is provided. Some experience from modern wars refer that AP mines do not provide any military efficiency. On the contrary, when war ends, demining is dangerous, slow and costly task, and land polluted with mines is a threat to lives of millions of people and cannot be cultivated.

Regarding their fuse type and function, AP mines types are: destructive-penetration, tripwire-bouncing (Bouncing Betty), blast-fragmentation and claymore mine. AT mines are destructive-penetration mines. All mine types could be used for booby-traps and/or with time delay function. In a war, mines are used as mine-explosive obstacles in order to cause casualties to the opponent, delay its advance and prevent use of certain corridors [3]. In such a way, mine became means of military defence in a first place. Because of technical abilities of quick remote placing, mines became military offensive weapons which can impede the opponent's defensive actions. Land mines are usually placed in minefields which can be AP, AT or combined minefields. AT mines are laid the fastest using mine-layer—embedded into dike, at distance of 4.5 m (5.0 m) [4]. However, mines are often placed individually as an obstacle (at the same time used as booby-trap mine) or in groups—without any rules of their placing geometry. AP mines, besides their use in AP combat, often have a role to protect AT mines, in order to prevent their removal. Mine suspicious zone is evaluated area where single mines and minefields can be found. Minefield is determined by geometry, shape, dimension and density, determined by military standards. Obstacle and mortality of minefield is evaluated by density—average number of mines per one meter of minefield width. Minefield density depends on its purpose, mine types, soil and tactical situation. Combined minefields, as well as booby-traps mines types, are very often used.

Obstacles can be features of the terrain that impede the mobility of a force. Some obstacles, such as mountains, rivers, railway embankments, and urban areas, exist before the onset of military operations. Military forces create other obstacles to support their operations. Commanders use these obstacles to support their scheme of maneuver [3]. When obstacles are integrated with maneuver and weapon fire, they can be decisive factor on the battlefield. For some obstacles, doctrine that relies on a physical object to impede vehicles or dismounted soldiers, such as antitank ditches, wire, road craters, and many types of roadblocks, has not changed since World War II. Because these obstacles do not damage or destroy equipment, or injure or kill soldiers, they are considered to be passive.

Although minefields are also obstacles, they are not passive, and doctrine for mine warfare has changed significantly. Today's mines are different from the mines of the World War II era, which required physical contact and relied on blast effects. Today's mines are triggered not only by pressure, but also by seismic, magnetic, or other advanced fuses.

The following operational concepts as possible alternatives to AP mines are considered:

- Use of mechanical ground systems, such as trip flares and improvised noisemakers.
- Use of electronic ground systems, such as the remotely monitored battlefield sensor system and ground-based portable radars.

Provide additional human systems, such as armed soldiers and more effective equipment, including binoculars, night-vision devices, or other capabilities; more deep-reconnaissance/surveillance units in conventional or special operations force organizations. Call upon airborne systems, such as unmanned aerial vehicles, helicopters, fixed-wing aircraft, joint surveillance target attack radar system, and satellites.

1.1.3 Demining Stress Factors

Humanitarian demining are activities which leads to the removal of mine and UXO hazards, including technical survey, mapping, marking, mine clearance, post-clearance documentation, community mine action liaison and the handover of cleared land. Demining may be carried out by different types of organization, such as NGOs, commercial companies, national mine action teams or military units.

Demining is a very stressful task for deminers, who perform extremely dangerous job. All mine polluted areas represent a *psychical area*, in which humans feel unsafe [5]. Entering such an area causes increase of psychical activity and attention. Conscious entrance of demining personnel in mine polluted area, with random mine placement, puts human sensor capabilities into state of increased awareness. Increased stress requires engagement of psychologist within demining teams, including personnel selection, preparations, demining projects monitoring, as well as psychological assistance after demining, especially after critical moments and incidents.

Demining stress factors are:

Environmental stress
Psychological condition
Defect of cohesion and psychosocial climate
Inadequate or lack of social support.

Environmental stress arise from psychological factors of demining tasks (such as perilous work conditions, boredom and monotony at demining, demining jammers and false alarms), climate and terrain demining conditions (heat, cold, humidity, area, noise and machine vibrations), and protective clothing factors and working equipment (protective and ergonomic factors of equipment).

Psychological state of deminers arise from his intellectual capability and personality, from questioning his qualifications, psychosomatic difficulties, inadequate stress interfacing, fatigue, medical condition and deminer's age.

Quality of interpersonal relations arises from level of confidence in co-workers and managers and situation within team after critical moments and incidents. Level of social support arises primarily from family relations, friend's support, public opinion on demining necessity and its promotion—media.

1.2 Humanitarian Demining Technology

Regarding different demining conditions, standard demining technology is combined technology, performed manually and by machines, and also using mine detectors and mine detecting dogs [4]. Combined technology provides methods and solutions for removal of different mine threats; speeds up the work process, and decreases risk and demining costs. In accordance with international mine action standards (IMAS) and demining conditions, national Mine Action Centres (MAC) and companies involved in demining develop standard operative procedures (SOP) for demining operations. Combined technology of the Croatian Mine Action Centre (CROMAC) is shown in the Fig. 1.1, whereas the different kinds of humanitarian demining equipment are in Fig. 1.2.

Fig. 1.1 Humanitarian demining technology (*Source* CROMAC, Ref. [13]) Demining capabilities ratio: 1 machine per 3 dogs or per 15 deminers

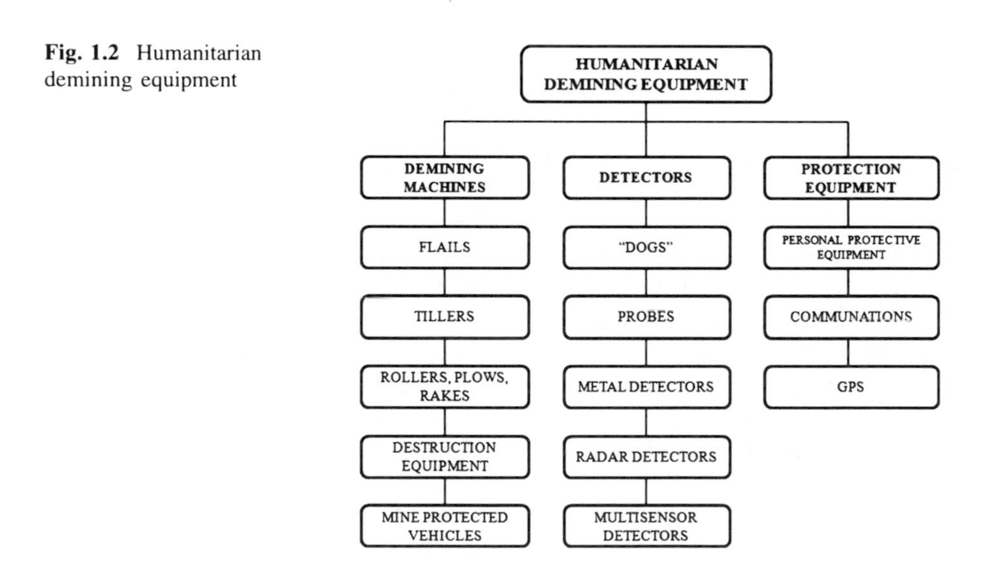

Fig. 1.2 Humanitarian demining equipment

1.2.1 Specification of Mine Clearance Quality

International mine action standard [6] defines "clearance" and specifies the quality system (i.e. the organization, procedures and responsibilities) necessary to determine that land has been cleared by the demining organization in accordance with its contractual obligations.

Land shall be accepted as "cleared" when the demining organization has ensured the removal and/or destruction of all mine and UXO hazards from the specified area to the specified depth.

The specified area to be cleared shall be determined by a technical survey or from other reliable information which establishes the extent of the mine and UXO hazard area.

Note: Priorities for clearance shall be determined by the impact on the individual community balanced against national infrastructure priorities.

Specified clearance depth shall be determined by a technical survey, or from other reliable information which establishes the depth of the mine and UXO hazards and an assessment of the intended land use. In the absence of reliable information on the depth of the local mine and UXO hazard, a default depth for clearance shall be established by the national mine action authority. It should be based on the technical threat from mines and UXO in the country and should also take into consideration the future use/purpose of cleared area.

In order to detect mine-explosive ordnance on mine suspected terrain, combined surveying method of human and machine terrain inspection is used. After terrain inspection demining machine is used as first demining method. It has a role to neutralize embedded mines by treating the soil. Machine has to remove the biggest risk from mines, and as such becomes the main equipment in demining mechanization.

Second method for terrain inspection, after machine, is deminers control, using metal detector or mine detecting dogs. Dogs are working in pairs; terrain is inspected with both dogs at the same time. Depending on terrain configuration and soil type, where machine cannot operate, deminers are processing the area and dogs are used for quality control.

Deminer is a person, including a public servant, qualified and employed to undertake demining activities or work on a demining worksite (*pyrotechnician, in Croatian*) [7].

The most effective demining technique in the world, at the present time, is detection of a metal signal with a World War II vintage metal detector, location of the landmine with a probe, and destruction of the landmine where it is found with an explosive charge. The major problem with this technique is twofold. First, the metal detector cannot tell the difference between a shell fragment, a piece of barbed wire, a soda can—common examples of *false positives,* and a landmine. The appropriate hand-held detection technology, which does not yet exist in a field—practical form in the world, must be able to determine the size, shape, material composition, and orientation of an object that causes a signal register in a detector.

To detect mines manually, modern transportable metal detectors are used. Non-metal—plastic AP and AT mines contain very small amounts of metal, so metal detectors detect them based on principle of electromagnetic induction. For example, at PMA-2 and PMA-3 mines, the only metal part of mine is thin aluminium fuse capsule which is a part of chemical fuse. Detectors detection depth is 15–20 cm. If mine is embedded deeper or is under vegetation, it is harder to detect the mine. Useful detection range/depth is limited by vegetation height, so vegetation has to be removed first. Under the influence of erosion and vegetation growth, mines are embedded deeper into the soil which makes detection more difficult. Although lots of false signals from metal fragments and metal scraps can be found, metal detectors are used, especially for detecting of mine obstacles. Radar testing that penetrates into the soil layer proves that deeper embedded mines can be detected. The goal for detector development is to develop a multi-sensor detector that combines performances of different sensors.

1.3 Efficiency of Demining Techniques

In order to develop demining mechanization, it is important to determine efficiency of existing demining technique in demining conditions. In line with the experience in humanitarian demining, deminer efficiency, demining machine testing and mine detecting dog testing, basis for evaluation of efficiency in humanitarian demining process is derived and set up by Croatian Mine Action Centre [8].

Based on knowledge of the Croatian Standards Institute it has issued a standard HRN 1142:2009 [9] which includes achievements in technique and technology of machine demining, methods and rules for demining as well as results of international standardization in countermine activities. The standard defines the technical requirements and procedures of evaluating machines used in humanitarian demining together with their acceptability. The background for setting up this standard was the CEN's document on testing and evaluation of demining machines, CWA 15044:2004 [10].

1.3.1 Classification of Demining Machines

Demining machines are classified as follows [9, 10]:

Classification according to mass

- Light, up to 5 tonnes
- Medium, 5–20 tonnes
- Heavy, more than 20 tonnes.

Fig. 1.3 Flail demining machine (*MineCat 230,* NoDeCo Norwegian; *Bozena 4,* WAY Industries, Slovakia; *MV-10,* Dok-Ing, Croatia) and tiller demining machine (*MV-4,* Dok-Ing Croatia; *MineWolf,* MineWolf Systems AG, Deutschland; *Mine Guzler,* Bofors, Sweden)

Classification according to mode of operation

- Direct operation from the cabin of the machine
- Operation with remote controls
- Operation with remote controls and video monitoring.

Classification based on tool

- Machine with flails
- Machine with tillers/mill
- Machine with rollers
- Machine with a vegetation cutter.

Selected representatives of flail machines and tiller machines are shown in Fig. 1.3.

1.3.1.1 Machine with Flails

According to the diversity of the ground clearing tasks, flails are designed in several categories, light, medium and heavy. Flails operate by counter-directional digging method down to 20 cm deep (flailing towards the shield) or co-directionally (ground clearing). Working tools for mine neutralization resist explosions of antipersonnel mines without being damaged, but anti-tank mine explosions can cause damage of the tools that will eventually require repair or replacement. The basic features of the flails are:

Light flails are designed with remote control, usually with tracked drive system. In front of the machine there is a flail made of chains and a hammers for digging, mine neutralization (crushing or activating) of underground or surface mines. Operating digging width is 1.5–2.0 m. It requires an engine power of 100–160 hp, and hydrostatic power transmission. The flail axle rotation speed is 500–1000 revolutions per minute. Operating speed of the machine is 0.5–1.5 km/h, depending on the working conditions.

Medium flails are designed on medium weight machines, tracked or wheeled, e.g. forest articulated tractor, 4×4 or 6×6 wheel drive. In front of the machine there is a flail for ground digging, crushing or activating of underground and surface antipersonnel and anti-tank mines. Flail engine power is 150–300 hp. Flail rotates at 500–900 revolutions per minute. Operating clearance scope is 2.0–3.0 m, and the operating speed of the machine is 0.5–2.0 km/h. Flails are operated by the engine operator directly from the cabin or by remote control.

Heavy flails are designed on the undercarriage of special vehicles and machines, such as forest articulated machines with high passing capacity 6×6 or 8×8. In front of the machine there is a driving unit of the flail. Engine power of the working unit amounts to around 500 hp and of the mobile vehicle about 200 hp. In the front it is fitted with flails with hammers for digging the ground and destroying of antipersonnel and anti-tank underground mines. Flail rotates at a speed of 300–600 revolutions per minute. Operating clearance scope is around 3.0–3.5 m, and the operating speed is 0.6–2.5 km/h. Heavy flails are operated by the engine operator.

Flail operation is based on the impact force of the flail (chained impact hammer of 0.5–1.5 kg), i.e. the impact moment of the flail. Depending on the humidity, soil is cut by the hammer blows and thrown towards the guard. If the soil is dry and hard, it is then crushed and dispersed towards the guard. Thus, there are two theories of soil digging, the theory of cutting the soil and the theory of crushing or strength of the soil. Therefore, there are two types of tools commonly used on machines. Cutting blade is used for the soft soil tool, and for the hard soil and crushing, rectangular form is preferred. The required force of the flail is calculated, in order to overcome cutting resistance and the soil strength resistance. Unlike tiller machines, flails are of simpler design. Distance of the flail rotor from the explosion centre equals the radius of the flail rotation. Because rotor and the impact hammer are connected through chain, in the ideal case the flail rotor is not subjected to the resistance momentum, therefore requiring less power for flail operation compared to the tiller machine operation. However, following the impact to the ground and pulling creates additional digging resistance, augmenting the resistance momentum.

1.3.1.2 Machine with Tillers

Machine with tillers as well as flails can dig the ground counter-directionally or co-directionally. The counter-directional ground digging is used for striking the mines "below the rotor" in order to smash them, while co-directional digging is used for throwing out or neutralizing mines "in front of the rotor". During soil digging, increased number of tool blades (or teeth) grips the ground, and that number depends on the digging depth. Lack of uniformity in the number of gripping tools, together with change in the ground cutting thickness, results in uneven digging resistance, including the necessary uneven rotor momentum. Digging the soil by crushing consists of the *cutting phase* and *layer removal phase*. Total resistance of the first phase consists of tangential resistance component (tangential to the blade path), and the perpendicular resistance component directed towards the rotor axis, i.e. the radial component. Greatest resistance acts on the tool blade that is leaving the soil grip. Shift resistance between the material and the tooth winding is more significant since the dug-out material is partly shifted along the rotor winding.

1.3.1.3 Supporting Demining Machines: Excavators

In complex projects, such as MSA on inaccessible terrains, adequate excavator mechanization and light demining machines are used. Hydraulic excavators with long arms and special working tool clear suspicious minefields where other machines cannot be used. Excavators are used in mine suspicious terrain such riverbanks, dikes, channels, soft soil, woods, intersected terrain, etc., Fig. 1.4. Thereby, excavators use one or two of special tools: demining flail/tiller or vegetation cutter. Based on previous experience, in complex demining projects, excavators from different manufacturers are used in combination with light demining machines. Light and medium demining machines are used for creating an approach path for excavators and its operation. Then, excavator is used to cut medium and high vegetation. If vegetation cutter is replaced by flail, excavator can demine suspicious areas. Often, after vegetation cutting light and medium machines that dig the soil and destroy embedded mines are used. On areas where light or medium demining machines could not be used, excavator is only machine that can perform demining operation by the use of flail. In these conditions

Fig. 1.4 Supporting machines—demining excavators (flail, tiller)

excavators have proven its quality in mine clearing on different terrains based on efficiency, precision, safety, cross-country mobility, crew safety and logistics. Advantages of demining excavator and flail type demining machine is in reduced safety risk for deminers with fewer people being involved, providing fast demining method and control.

Tracked or wheeled excavators of 15–30 tons are being modernized in accordance with user's needs in real demining terrains. Clearing width of excavator's flail is 1.0–1.5 meters. Excavator arm can be extended for increased demining access (long range, 15 m), increasing excavator's working area. Hydraulic lines are placed along the arm and protected against explosion fragments. Additionally, armour protects the excavator and regular driver's cabin is being replaced with an armoured one. Driver's safety is increased—being protected against shrapnel and AT mine explosion blast and noise, as well as against rolling and falling objects.

As demining technology is constantly being improved, demining machine manufacturers are improving and modernizing their products in accordance with demining technology requirements. Modern technology of excavator's design was proven in testing on proving grounds as well as in real environment. Regarding efficiency, special excavators provide increased efficiency in complex conditions. Efficiency of such excavators is 500–1000 m^2/h. Regarding demining, companies competing on the market are modifying their own excavators or use excavators on lease. Excavators demining efficiency in complex demining projects is obvious, especially in extremely difficult conditions and when demining costs per m^2 are the highest. In the world of humanitarian demining such special excavators are more and more present. For companies dealing with demining, is affordable to buy excavators and adapt them to their need, developing special demining tools. In such conditions, excavators will soon pay off, probably sooner than other demining machines.

1.3.2 Demining Conditions

Conditions defined by soil type and category, terrain configuration and vegetation are referred as demining conditions [9]:

The following demining conditions are differentiated:

a. Favourable soil conditions
b. Aggravated soil conditions
c. Difficult soil conditions
d. Specific conditions.

When describing soil conditions apart from soil categorization, it is also necessary to take into consideration the following: terrain inclination, soil humidity as well as vegetation and its categorization.

According to the terrain inclination, there are:

a. Flat soil: possible longitudinal and transversal inclination 0–5°,
b. Inclined soil: possible longitudinal and transversal inclination 6–15°,
c. Steep soil inclinations: slopes of canals, ditches, dams etc.

According to the level of soil moisture there are:

a. Dry soil,
b. Soil with increased level of moisture,
c. Swampy soil.

According to the vegetation there are:

a. Low vegetation: up to 1 m high, mostly annual underbrush, of higher or lower density, with the existence of rare low bushes,
b. Medium vegetation: up to 1–2 m high, with the existence of bushes and single trees up to 10 cm in diameter,
c. High vegetation: over 2 m high, with the existence of bushes and trees over 10 cm in diameter,
d. Forrest: trees over 3 m high, over 10 cm in diameter.

1.3.2.1 Favourable Soil Conditions

Favourable soil conditions are characterized by flat soil with possible longitudinal and transversal inclinations from 0 to 5°. Soil is mostly humus, loam or sand of normal (medium) hardness and low level of humidity, covered in low vegetation. Low vegetation consists of fresh or dry grass, of higher or lower density, weeds, rare low bushes up to 0.5 m high. Soil is easily treated with manual tools (shovel and pickaxe) and stipulated use of probe is possible (penetration into the soil up to the planned soil treating depth) as well as metal detector. There is no mineralization of soil or soil contamination with metal.

1.3.2.2 Aggravated Soil Conditions

Aggravated soil conditions are characterized by hilly terrain with possible longitudinal and transversal inclinations from 6 to 15°. Soil is mostly humus, loam or sand of increased hardness (very dry) or increased level of humidity, dirt mixed with rocks, dirt prevails with rare low and medium vegetation. Stone is limestone-schist, soft, easily crushed by machine working tool. Low vegetation is up to 1 m high and medium up to 2 m. There are some rare trees up to 10 cm in diameter. Soil is treated with more difficulty by the manual tool for the reason of increased hardness of soil (by shovel and pickaxe) and probe is also used with more difficulty (it is hard to achieve penetration up to the planned soil treating depth) as well as metal detector for the reason of present soil mineralization and metals. Soils with increased level of humidity and underwater soils are also included into aggravated soil conditions.

1.3.2.3 Difficulties Soil Conditions

Difficult soil conditions are characterized by soil with steep slopes of canals, ditches dams and bigger hills. Soil is mostly stony, stone sheets with dirt between them, possible vegetation existence. Stone is of medium hardness, use of machine working tool is possible on reduced depths (up to 10 cm). Present vegetation is underbrush, bushes of high density over 2 m. There are single trees over 20 cm in diameter. Forest areas are criss-crossed with ditches and water-worn ravines, trees over 3 m high and over 20 cm in diameter. Soil is treated with difficulty by manual tool for the reason of increased soil hardness and presence of stone, stipulated use of probe is possible only in certain places. High level of soil mineralization is present as well as soil contamination with metal. Swampy soils with and without vegetation are included into difficult soil conditions. If it is about extremely underwater and swampy soil, use of demining machine is limited.

1.3.2.4 Specific Conditions

Specific conditions are characterized by soil that is difficult to describe with already mentioned above conditions and defined categories of soil and vegetation. These are very steep chains of mountains, river canyons with big rugged slopes, thick and impassable woods, impassable rocky ground overgrown in thick underbrush or single thick small forest with trees over 3 m in height and over 20 cm in diameter. Soil type makes it very difficult or even impossible to use demining machine with acceptable results. It is not possible to treat soil with

Table 1.1 Efficiency of deminers using manual mine detection—primary examination method in different soil and vegetation conditions (*Source* Ref. [8])

No.	Site conditions		Evaluation
	Soil condition	Mine condition	Efficiency (m^2/day)
1.	Difficult and Specific	Confirmed minefield existence	to 50
		Assumed minefield existence	to 75
		No indication of minefield existence	to 150
2.	Favourable and Aggravated	Confirmed minefield existence	to 75
		Assumed minefield existence	to 150
		No indication of minefield existence	to 250

Table 1.2 Estimate of deminer's efficiency using combined method—combination of demining machines and manual mine detection (*Source* Ref. [8])

No.	Soil treatment depth	Efficiency (m^2/day)
1	Manual detection after the use of demining machines digging at depth required by the project	450
2	Manual detection after the use of demining machines that could not digging at depth required by the project	300
3	Manual detection after the use of supporting machines used for area survey and demining	250

manual tool because of higher soil hardness and presence of stones, stipulated use of probe is not possible in acceptable way. There are ruins and imploded storage buildings with explosive remains of war, chasms and wells with different ordnance, scattered rubble etc.

1.3.3 Efficiency of Deminers

Manual mine detection as a primary method and as a secondary method
Efficiency of deminers using manual mine detection is given in Table 1.1 and an estimate of deminer's efficiency using combined method is given in Table 1.2.

Table 1.3 Estimate of mine detection dog's efficiency using the combined method—using dog—handler team (*Source* Ref. [8])

No.	Work method of dog—handler team	Efficiency of the team[a] (m^2/day)
1	As a second method after manual mine detection	1,500
2	As a second method after the use of demining machines digging at required depth	1,500
3	For inspection of ruins and/or destroyed buildings and stratified removal of material sediments	500
4	For final demining quality control	500

[a] average values

1.3.4 Efficiency of Mine Detection Dogs

When performing technical surveillance, mine search and demining activities in mine suspected area (MSA), mine detection dogs could be used as a second method (engaging dog—handler team) in combination with the manual mine detection method (metal detector and probe) or in combination with demining machines, and also as a primary method for searching the ruins at stratified sediments removal. An estimate of mine detection dog's efficiency using the combined method is given in Table 1.3.

1.3.5 Efficiency of Demining Machines

It is apparent that demining machine efficiency is lower on a hilly terrain, particularly in difficult conditions. Basic types of demining machines are shown in Fig. 1.5 and an estimate of their efficiency in Table 1.4, i.e. shown in Fig. 1.6. The greatest efficiency is provided by heavy machines on flat terrain. An estimate of supporting machines efficiency—excavators is given in Table 1.5. It can be seen that tracked excavators with long reach have the greatest efficiency.

It is obvious that machine demining provides the best efficiency in humanitarian demining. Depending on machine category and working conditions, light machines should provide efficiency of 500–1,000 m^2/h, medium machines 1,000–2,000 m^2/h, and heavy machines 1,500–2,500 m^2/h.

Deminer's efficiency, when using manual mine detection method in favourable conditions should be two times higher, if demining machine is used first—450 m^2/day.

Dog—handler team efficiency, after the use of demining machine, is five times higher than deminer's—1,500 m^2/day.

In comparison with manual demining, mechanization of demining significantly decreased total time for demining and increased deminer's safety.

Fig. 1.5 Basic categories of demining machines with flails; Light machine (*above*), Medium machine (*middle*) and Heavy machine (*below*)

1.3.5.1 Demining Knowledge and Develop Demining Equipment

If a country wants to demine large area quickly, it needs to develop demining knowledge and develop equipment for machine demining, and set up standards of humanitarian demining. An example of this is the standard of humanitarian demining—Requirements for machines and conformity assessment for machines, issued by the Croatian Standards Institute.

One differentiates between mine danger awareness and solution, as seen in Fig. 1.7. Mechanization procurement costs can be lowered if MSA characteristics are known, in order to choose adequate demining machines and other equipment. Selective and combined use of machines and equipment for demining projects for real demining projects provides optimal solutions.

Table 1.4 Estimate of demining machine efficiency in real conditions (*Source* Ref. [8])

Machine category	Working tool	Efficiency in mine suspected area (m^2/h)			
		Flat terrain		Hilly terrain	
		Favourable conditions	Aggravated conditions	Favourable conditions	Aggravated conditions
1. Light machine	Flail	650	450	500	400
	Flail	800	600	650	500
Average		725	525	575	450
2. Medium machine	Flail	850	650	600	500
	Flail	1,800	1,650	1,500	1,000
	Flail (double)	2,000	1,900	1,850	1,450
Average		1,550	1,400	1,300	980
3. Heavy machine	Flail (double)	2,600	2,000	2,000	1,800
	Tiller (double)	2,400	1,800	1,800	1,600
Average		2,500	1,900	1,900	1,700

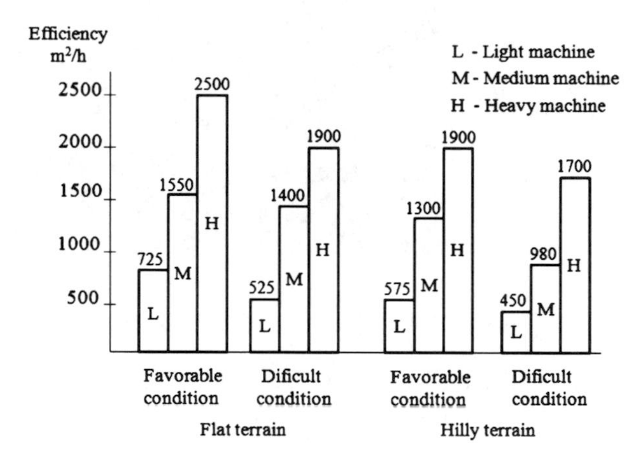

Fig. 1.6 Estimate of average demining machine efficiency

1.3.6 Experience of the Machine Demining

1.3.6.1 Preparation for Machine Demining

Site position scheme of control and logistic services is shown in Fig. 1.8. Preparation for machine demining includes: preparation of properly cleaned access

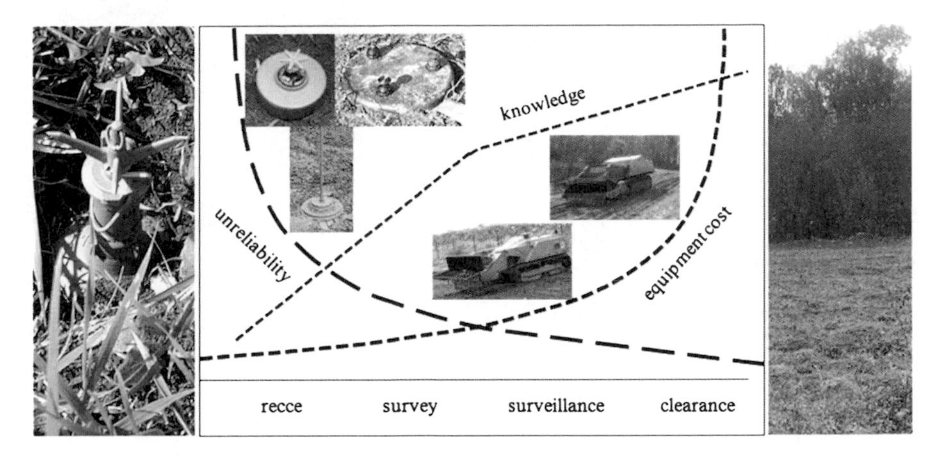

Fig. 1.7 Diagram mine danger awareness and means of solution

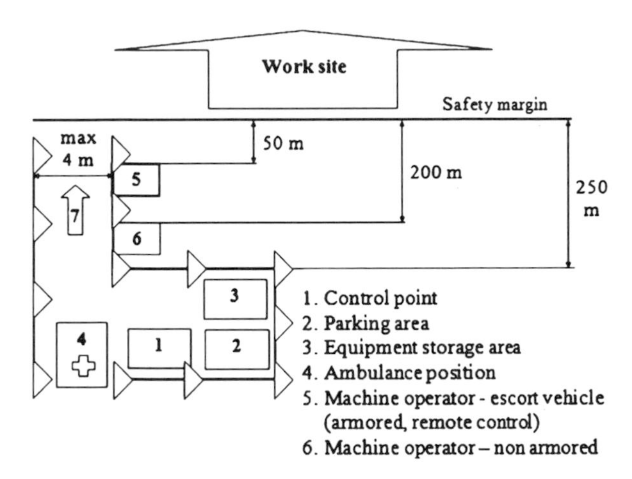

Fig. 1.8 Site position scheme of control and logistic services

Table 1.5 Estimate of supporting machines efficiency—excavator (*Source* Ref. [8])

Machine	Working tool[a]	Efficiency[b] (m^2/h)
Excavator—wheeled	Flail	250
Excavator—wheeled, long range	Flail	400
Excavator—tracked, long range	Flail	500
Excavator—tracked, long range	Flail	650
Excavator—tracked, long range	Flail	750

[a] for removing high vegetation, vegetation cutter is used, and then demining flail; working tool width is 1.0–1.5 m

[b] excavator efficiency in favourable conditions—flat terrain

path of up to 5 m width, secure the safety lane and turning area, overlapping with neighbouring lanes, and place for visual inspection of machine operation, as in Fig. 1.9 and Fig. 1.10. Turning area for heavy demining machines should be at

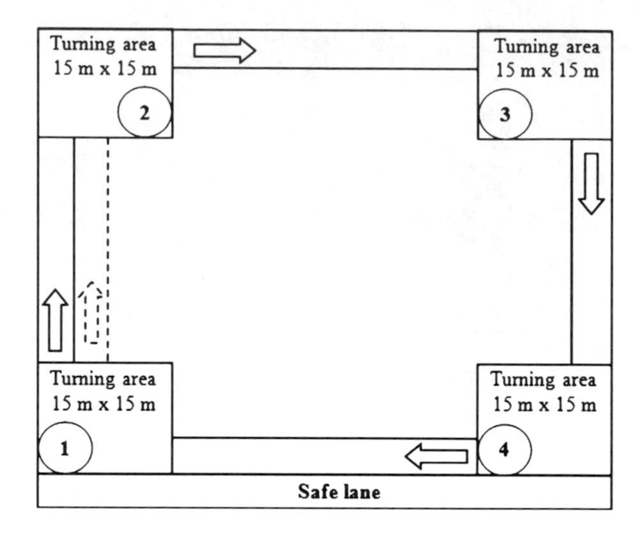

Fig. 1.9 Scheme of safety lane and machine turning area

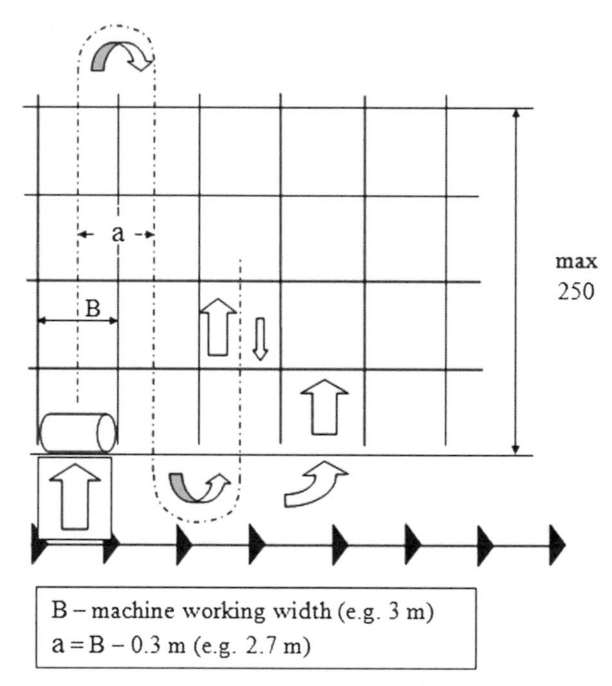

Fig. 1.10 Scheme of maps and referent points for lane inspection in coordinate grid

least of 15 × 15 m in size. Starting points and demining lanes begin from safety zone or turning area, at the angle of 90° if possible. If it is not possible to build up turning area in the corners on the opposite side of safety lane, machine operates up to demining boundaries and moves backward on the same track. At least three passes should be made, before continuing to the next corner of demining area. After that, operation continues in circular movement until the area in the middle

Fig. 1.11 Heavy demining machine (RHINO 02, MaK System GmbH)

decreases. Demining lanes can be of 1000 m in length. For machine demining, vegetation should be removed. Before machine demining, medium and higher vegetation should be removed with machines.

In case of machine failure or damage, it is important to enable its towing out of minefield. For this reason, some machines are designed with two engines, so that other working engine can move the machine out of the "mouse" trap using remote control. In case of working tool failure, machine is towed out from the minefield. If machine is stuck or its drive fails, i.e. if machine cannot pull itself out, safety lane towards machine is prepared manually, as well as safety lane around machine of up to 4 m in width, in order to prepare machine for self-towing.

Soil clearance control after machine demining is performed with metal detectors or dogs. Faster and cheaper method is use of machine detection method. Control using metal detectors is done according to SOP. After that, mine detection dogs are used for control of explosive remains. It is important to be familiar with possibilities for use of dogs. After machine demining, several days should pass before the use of mine detection dogs on treated area. That period will enable dissipation of scent from crushed mines, as well as to scatter the scent from leaked fuels and oils. Explosions at machine demining leave detonation products in the soil, which can confuse the dogs. If there were no explosions, dogs are used for clearance confirmation after machine demining operation.

1.3.6.2 Clearing Records of the Heavy Demining Machine

Mine clearance by machines is the safest tool for preparation of mine clearing areas for later safe/control check using metal detectors for metal parts of mines and dogs for explosives. The job of mine detection is much faster, easier and of course safer, if it is performed in an already treated area.

Consequently, after clearing over 10 square kilometres with machines, all types of AP mines, like PROM-1—the most dangerous bouncing AP mine, were destroyed in a safe way for the deminer, even without damage to the machine. AT

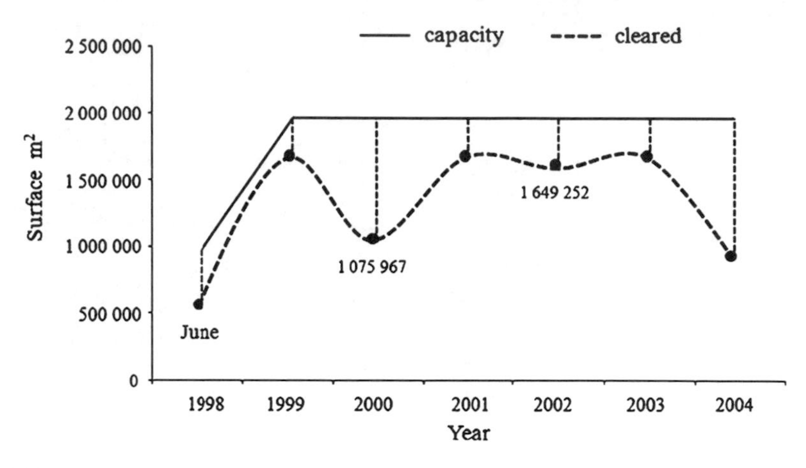

Fig. 1.12 Clearing records of the heavy demining machine RHINO 02 (Reproduced with permission from Ref. [11])

mine is usually causing some damage that can be repaired in the field in a short time the same day. In general, AT-mine damage do not happen very often, because some mines will be crushed, and the operator can see the remaining part using remotely controlled heavy demining machine (*RHINO 02*) equipped with color video camera, Fig. 1.11. For the maximum quality, it is essential to have a clearing system with adjustable automatic depth control. Typical and most used clearing depth is 20–30 cm, this is unachievable result for demining with a probes and/or a metal detector.

Efficiency of a heavy demining machine *RHINO 02* 40 t [11] in the period of 6 operating years is shown in Fig. 1.12. Under real conditions of machine engagement, mine clearance efficiency amounts to over 1,500,000 m^2.

Compared to manual demining, machines are also operating fast on rocky areas, areas with dense vegetation, covered with grass, bushes and small trees. Furthermore, this ensures that trip wires and most mines are removed for a much safer second check for metal and explosive parts of mines. Additional advantage of machine clearing is that the cleared areas are excellently prepared for agricultural purposes. There are machines, that can work with larger and denser vegetations, like bigger trees and woods, but this is not a practice because it is slowing down the clearance and makes it more expensive. Safety of mine clearing can also be achieved by the use of machines. An underlying precept of successful demining operation is a perfect safety record for demining personal and the users of cleared areas.

Safety for machine operators requires either remote control with audio and video surveillance or alternatively an excellent protection for the driver cabin/cockpit. Remotely controlled equipment is better solution, due to possible driver's psychological stress because of being exposed to explosions, even though he is protected by armoured cockpit.

Safety for deminers means much easier and safer use of metal detector or probes or dogs after the area is cleared by a machine. Possibility of accident and stress factors is reduced. Safety for users of the cleared area is ensured through quality of mine clearing.

Based on records of *RHINO 02* machine, the overall cost of the demining in operation was reduced to 2/3 in comparison to strictly manual demining. For example, if there is a plan to clear an area of 200 ha in a year (this is equivalent to an area of more than 400 football fields), it is more cost effective to use combination of machines and a reduced amount of deminers compared to the strictly manual demining:

- It lowers all costs by 1/3
- It reduces personnel by 1/3
- Provides for maximum safety and quality level

Personnel for demining with machines	Personnel for manual demining
Deminers 8	Deminers 35
Medics 2	Medics 2
Mechanics 8	Others 3
Dog handlers (× 2 dogs) 4	Total: 40
Others 4	
Total: 26	

Combination of demining with machines, and an additional second check by deminers with metal detectors or dogs after machines is the best way of demining concerning quality, safety and costs.

1.3.6.3 Follow-on Processes After the Use of Demining Machines

In order to achieve efficiency in the established humanitarian demining quality-assurance/quality-control system, acceptable quality levels have been defined for mechanical ground processing using demining machines, manual-detection methods and/or a combination of the manual method and the use of mine-detection dogs.

Quality mine and UXO clearance ensures that the beneficiaries of cleared lands can enjoy a safe livelihood in the demined area. This level of confidence requires a high degree of organization and management, maximum engagement of demining organizations and national mine-action bodies, and a quality-control system with detailed operating procedures. Such a system guarantees the best supervision and control of the suspected mined area.

The Croatian Mine Action Centre has developed a quality-assurance/quality-control system QA/QC system [12], which entails the following three phases:

1. Quality Control
2. Quality Assurance
3. Total Quality Management

This QA/QC system follows the guidelines set in the following documents: ISO 9000, ISO 2859, IMAS 09.10 and IMAS 09.20. The International Mine Action Standards are an integral part of the management process aimed at verifying demining quality and reaching sufficient confidence (with acceptable quality levels) that the demining company has removed and/or destroyed all mines and unexploded ordnance from the mined area according to specifications.

1.3.7 Role of Machine Demining

An important question is to what extent machine demining participates in the overall mine clearance activities in the world, and the answer is of importance as well. It is crucial to provide the characteristics of advanced machine demining, where the share of machine demining is dominant. Important to note here is that the use of post-machine demining control methods according to the humanitarian demining technology is unquestionable (detectors, dogs, probes). There exists a very small percentage of mine-suspected areas where machines could not have been used.

Manual demining method includes familiarization with: procedures for use of probes, procedures for tripwire detection, procedures for use of metal detectors, procedures for vegetation removal, procedures for demining with explosives, procedures for mine detection, and procedures for mine destruction on site.

1.3.7.1 Application of Demining Machines

Demining machines can be use for:

1. deminer's inspection
2. deminer's survey
3. demining

Application of machines in deminer's inspection:

- machine enters into deminer's inspection area
- machine treats up to 10 % of MSA
- machine sets up the boundary lines of deminer's inspection area

Application of machines in deminer's survey:

- machine enters into deminer's survey area
- machine treats the whole surface of deminer's area
- machine destroys detected mines

Table 1.6 Fleet of demining machines CROMAC 2001–2011

Demining machines	2001	2003	2005	2007	2009	2011
Light, to 5	10	22	29	18	19	18
Medium 5–20	3	4	8	14	21	17
Heavy >20	4	4	7	8	8	12
Excavators >15	5	10	15	6	7	8
Total	22	40	59	46	55	55

(*Source* Ref. [14])

- upon machine treatment, at least 15 % of deminer's survey is performed using other methods
- if systematic mine pollution is identified in certain area by the use of machine, that area is categorized as "area that has to be demined"

Application of machines in demining projects:

- machine treats the whole area
- machine destroys mines by activating or crushing (neutralization)
- upon machine, use of other methods are mandatory (manual method, detection dogs)

1.3.7.2 The Advantages of Machine Demining

The advantages of machine demining is in reduction of safety risks for the deminers with fewer people being involved, providing fast demining method, i.e. higher performance and high level of mine clearance. The advances in machine technology regarding work safety and improved efficiency in clearing of the smallest mine remains in all terrain conditions ensures the future of demining machine technology. Strict requirements are being set regarding efficiency and economy of the machines. New solutions of demining devices are being tested. Such devices include flails and milling machines (mill/tiller) as basic devices in humanitarian mine clearance. The demining companies prefer flails, as a simpler and less expensive solution, and because these are more efficient, it ensures higher profit.

[1] Organised humanitarian mine clearance of the mine contaminated areas of the Republic of Croatia began with the foundation of the Croatian Mine Action Centre (CROMAC) in 1998. In that period noticeable advances in the theory and practice of machine demining have been achieved. Machine demining is becoming increasingly present in the mine clearance activities. Demining machines have contributed to higher performance, improved safety and quality of work as well as lower price of demining. This is the result of understanding their role and development for the actual demining conditions. Thus, in 2003, the percentage of machine demining amounted to 85 % of the treated area. Machine demining considers the prime role of machines, upon which other detection methods for mine clearance control can be used.

According to the CROMAC[1] data given per Table 1.6, a growth in the number of demining machines is evident during the time period. In the year 2003, the 40 machines had cleared 24 km^2, which is 85 % of treated surface. Proportion is: light machines 53 %, medium machines 10 %, heavy machines 10 %, and excavators 27 %. During the year 2005 number of machines was increased to 59. During 2007/2009/2011 the number of machines is maintained around a constant of 55 accredited machines. True machine efficiency depends on awarded contracts and machine engagement in these projects.

Accordingly, equipping of companies is based on all demining machine types. It is noticeable that percentage of light machines and excavators is increasing from 2001 to 2005. From 2007 to 2011 the number of accredited machines is approximately held constant. However, the number of light machines has decreased in the same period, whereas the medium and heavy machines tended to increase. This is explained by the fact that the light machines were used to demine rural areas, which are of priority interest. The rest of mine-suspected area applies to large surfaces where medium and heavy machines are used. This equipping concept can be considered as optimal for small and medium size companies. Machines can be specially developed or other machine adapted for humanitarian demining. Such a dual source of demining machines is the basis for equipping of demining companies, in order to be able to respond to future challenges. Different machines combined on one working site, in relation to working conditions, can provide maximum demining efficiency—providing for lower costs, higher working speed and better deminer safety.

References

1. Zunec O (1997) Planet Mina (The Mine Planet), Strata Investigations, Zagreb.
2. Landmine Monitor Report (2005) Toward a Mine-free Word, International Campaign to Ban Landmines, Landmine Monitor Editorial Board, Mines Action Canada, Handicap International Human Rights Watch, Norwegian people's Aid.
3. Croll M (1998) The History of Landmines, Leo Cooper, United Kingdom.
4. Mikulic D (1999) Demining techniques, modern methods and equipment, Demining Machines (in Croatian), Sisak, Zagreb.
5. Stefan S (2004) Psychological support for mine-clearing personnel in the Croatian armed forces, International symposium "Humanitarian Demining 2004", Sibenik.
6. International Mine Action Standard—Clearance requirements (2003) IMAS 09.10, 2nd Edition, United Nations Mine Action Service (UN MAS) New York.
7. Law on Humanitarian Demining (2005) Croatian Parliament, Zagreb.
8. Standard operating procedures SOP (2007) Estimating the effects of searching and demining, Possible efficacy of demining machines, Quality monitoring and control, Croatian Mine Action Centre, CROMAC, Sisak.
9. Humanitarian demining-Requirements for machines and conformity assessment for machines (2009) Standard HRN 1142, Croatian Standards Institute, HZN 1/2010, Zagreb.
10. Test and evaluation of demining machines (2009) CEN Workshop Agreement, CWA 15044:2009, Supersedes CWA 15044:2004, CEN, Brussels.

11. Druzijanic D (2005) Experience from Over 10 Square Kilometres of Mechanical Mine Clearing, International symposium, "Humanitarian Demining 2005", Book of Papers, Sibenik.
12. Gambiroza N (2010) The CROMAC QA/QC System, The Journal of ERW and Mine Action, Issue 14.1, CISR, James Madison University, Harrisonburg, VA.
13. Characteristics and technology of humanitarian demining (2006) Report, Scientific Council of CROMAC, Zagreb.
14. Croatian scientists' 10 years of work on the issues of demining (2008) CROMAC-CTDT, Zagreb.

Chapter 2
Mechanics of Machine Demining

Abbreviations

a_t [m/s^2]	Tangential hammer acceleration
b [m]	Tool blade width, tooth
CI [Pa]	Soil cone index
D [m]	Diameter of wheel
f_k [-]	Rolling resistance coefficient
F_a [N]	Mine activation force
F_i [N]	Hammer force impulse
F_{in} [N]	Grasping hammers force impulse
F_{cf} [N]	Flail centrifugal force
F [N]	Hammer striking force
F_N [N]	Striking force normal component
F_H [N]	Striking force horizontal component
F_1 [N]	Impact force
F_2 [N]	Dragging force
h [m]	Soil cutting depth
h_r [m]	Rotor axis height
k_1 [N/m^2]	Specific cutting resistance
k [-]	Collision factor
k_o [-]	Relative soil resistance
L [m]	Cutting tool width
M [-]	Mathematical expectation
MMP	[Pa] Mean Maximum Pressure
MI [-]	Mobility Index
m_h [kg]	Hammer mass
m_c [kg]	Chain mass [kg]
M_u [Nm]	Total flail rotation resistance moment
M_e [Nm]	Moment of eccentric rotor weight
M_μ [Nm]	Friction moment in shaft bearing
M_d [Nm]	Moment of flail to soil friction

D. Mikulic, *Design of Demining Machines*,
DOI: 10.1007/978-1-4471-4504-2_2,

M_i [Nm] Moment of shaft, hammer and chain inertia
NGP [Pa] Nominal Ground Pressure
p_i [Pa] Tyre inflation pressure
p_1 [%] Probability that disk sections will hit the mine in minefield
p_2 [%] Probability that mine will be activated
P_v [W] Power required for machine movement
P_r [W] Power required for machine operation (flail)
P_T [W] Total power required for machine movement and operation
P_m [W] Engine power of machine
r [m] Hammer rotation radius
R [%] Reliability of mine activation
R_1 [N] Cutting resistance
$R_{\sigma i}$ [N] Non-coherent soil resistance to crushing
R_{ki} [N] Coherent soil resistance to digging
R_k [N] Rolling resistance (wheels/tracks)
R_i [N] Inertia resistance
R_α [N] Slope resistance
ΣR [N] Movement resistance
S_t [m] Current cutting layer thickness
Δt [s] Time interval of hammer soil grasping
t [s] Acceleration time
u [m/s] Circumferential hammer velocity after collision
v_o [m/s] Circumferential hammer velocity before collision
VCI [Pa] Vehicle Cone Index
w [%] Moisture content
W [N] Wheel load
Z [m] Sinkage
ω [s^{-1}] Angular hammer velocity
θ [°] Hammer striking angle (angle of chain)
φ [°] Flail angle
ε [s^{-2}] Angular hammer acceleration
$_w$ [-] Wedge efficiency of striking hammer
α [°] Cutting angle
β [°] Wedge angle
γ [°] Back pin angle

2.1 Soil Categorization

Machine working conditions are very important for demining operations. It is necessary to be familiar with soil characteristics as well as mine features, because of their influence on machine design and demining technology. Soil characteristics, possible soil conditions and physical–mechanical characteristics are described by *Soil mechanics*.

Categorization of real demining conditions is extremely important, when demining equipment is concerned (probes, detectors, machines). Four soil categories where mines can be neutralized with demining equipment are presented in Table 2.1. Additionally, soil configuration is important for use of certain demining technique; flat soil with rocks may be on one location, hilly terrain with rocks may be on another location, and flat terrains on a third.

With regard to difficult treatment of soil with manual tool and demining machines, soil categorization and vegetation categorization [1] are provided, Tables 2.1, 2.2.

2.2 Soil Trafficability

Use of machines and vehicles in demining depends on state of soil. On soft and moist soil, machines and vehicles move with difficulty. There exists a problem of soil trafficability. This problem is more evident with wheeled vehicles than the tracked ones. Indicators of soil trafficability on basis of vehicle pressure on soft soil, are:

1. Nominal Ground Pressure (NGP)
2. Mean Maximum Pressure (MMP)
3. Soil load capacity, soil cone index (CI)
4. Rut depth (Z), sinkage
5. Vehicle Cone Index (VCI)

2.2.1 Mean Maximum Pressure

In a study of soil trafficability often times a nominal vehicle pressure is used on soil NGP (Nominal Ground Pressure), as an easiest approach to soil trafficability estimate. However, nominal pressure on soil is a marginal tangential pressure of wheel on soil, which doesn't provide a competent soil trafficability estimate because of neglecting the impact of laden wheel pneumatics deformation while moving, or because of track chain deformation.

Mean maximum pressure (MMP) is a referent pressure of the vehicle on soft soil through the wheels. It is defined as the mean value of peak pressure magnitudes acting on the soil under the wheels. A partial empiric model (*British Army Engineer Corps*) for evaluation of vehicle mobility on coherent (clay) soil has been developed. Lower value of MMP decreases wheels sinkage, which provides better soil trafficability and mean movement speed, which further provides better vehicle mobility. Linking MMP and CI soil load, correlation that defines vehicle mobility is being determined (go/no go). MMP Eq. (2.1) enables analysis of design factor

Table 2.1 Soil categorization

Soil category	Soil features	Method used
I	Medium and hard soil, covered with vegetation, humus, loam, compact sand	Manual tool, probes, shovel, use of machine.
II	Dirt mixed with rocks, dirt prevails with rare low and medium vegetation. Stone is limestone-schist, soft, easily crushed by machine working tool.	Probe used with difficulties. Use of machine.
III	Stony soil, stone sheets with dirt between them, low vegetation. Swampy soil.	Probe used on surface. Possible use of machine.
IV	Specific conditions, very hard soil, other categories not applicable.	Not possible to use probes. Machine is used with difficulties.

(*Source* Ref. [1])

Table 2.2 Vegetation categorization

Vegetation categorization	Vegetation features	Vegetation texture	Height and diameter
Low	Fresh or dry grass of low or higher density, weed, rare low bushes	80 % grass, and 20 % bushes	1 m
Medium	Grass, weed, single bushes, thick vegetation, single trees	50 % grass, 50 % bushes, 10–15 trees	1–2 m >ø10 cm
High	Bushes, weed, grass, high vegetation density, single trees	20 % grass, 40 % bushes, 40 % underbrush, over 15 trees	> 2 m >ø 10 cm
Forest	High forest and dense underbrush	High trees Tree diameter	>3 m >ø 20 cm

(*Source* Ref. no. [1])

influence on decreasing the pressure to the base and accordingly evaluation of mobility of different vehicles.

2.2.1.1 Wheeled Vehicle

$$\boldsymbol{MMP} = \frac{\boldsymbol{kW_T}}{\boldsymbol{nb^{0.85}d^{1.15}}\sqrt{\frac{\delta}{d}}} \; [\mathbf{kPa}] \tag{2.1}$$

W_T vehicle weight (kN)

k number of drive axis factor (2.05 for 4 drive axlex; 1.5 for 3 drive axlex; 1.83 for 2 drive axlex)

n number of vehicle wheels
b tyre width, non-loaded wheel (m)
d tyre diameter, non-loaded wheel (m)
δ tyre deflection due to the load (m)

Pneumatics deflection

$$\delta = \left(0.365 + \frac{170}{p_i}\right)\frac{W}{1000} \text{ [m]} \tag{2.2}$$

p_i tyre inflation pressure (kPa)
W wheel load (kN)

2.2.1.2 Tracked Vehicle

Use of machines in demining depends on the state of soil. When weather conditions change (rain, mud), soil load capacity (soil strength) also changes. When soil is humid and the machine is too heavy, it won't move or perform demining. Machines with less pressure on soil offer greater soil trafficability.

$$MMP = \frac{1.26\, W_T}{2nb\sqrt{Dt}} \text{ [kPa]} \tag{2.3}$$

W_T vehicle weight (kN)
n number of wheels of one track
b track width (m)
D diameter of wheel (m)
t track pitch (m)
L track length, in contact with level ground (m)

Nominal ground pressure

$$NGP = \frac{W_T}{2Lb} \text{ [kPa]} \tag{2.4}$$

2.2.2 Soil Cone Index

CI soil cone index is a soil load capacity indicator, which is measured with penetrometer, Fig. 2.1. Resistance to penetration of penetrometer cone into the certain type of soil is measured. *Standardized value of cone penetration measurement on depth of 15 cm (ASAE EP542 1999) is called cone index (CI).* For

Fig. 2.1 Penetrometer for measuring the soil cone index

example, load of very soft soil has soil cone index of $CI < 300$ kPa, medium hard soil is 300–500 kPa, and very hard soil more than 500 kPa.

Coherent soil load capacity [3], cone index CI:

0–21,	load capacity has no practical value
40–62,	a man has difficulty walking on soil without sinking
103–165,	special light tracked vehicles can surmount approximately 50 driveways
186–228,	light tracked vehicles can pass about 50 times
276–352,	medium weight tracked vehicles can surmount approximately 50 passages
372–497,	vehicles of Jeep type, can pass around 50 times
517–662,	heavy vehicles
683–935,	passenger vehicles
1000,	without problems in soil trafficability

Limiting cone index

$$CI = CI_L = 0.827\,MMP \ \text{[kPa]} \tag{2.5}$$

This expression can be used to determine the lowest soil load, where vehicle with certain MMP is mobile. In another words, soil load should be at least 83 % of given MMP for certain vehicle, in order to successfully cross the passage.

For multiple passes on the same coherent soil track and determination of adequate CI_L, multiplication index are used. They are gained experimentally [2] and given as *RI* coefficient of soil load alteration (*Remoulding Index RI*) which considers soil sensitivity regarding the soil strength loss due to road traffic. Evaluation of such cone index rate *(Rating Cone Index)* is $RCI = RI \times CI$.

Multipass multiplicator

Number of passes	1	2	5	10	25	50
RI	1	1.2	1.53	1.85	2.35	2.8

2.2.3 *Wheel Rut Depth*

Many factors effect vehicle moving, firstly its physical characteristics (structure, density, moisture, watertightness, and other) and mechanical characteristics (load capacity, shear resistance, cohesion characteristics, tackiness, and other). Coherent soil, while wheel rotation, plastically deforms. Under the wheel, soil is exposed to complex normal and tangential constraints. That leads to soil compaction and shear. Deformation degree depends on soil properties, i.e. its mechanical features. When the resistance for deforming is higher, soil provides more convenient conditions for vehicle movement, that is better soil trafficability.

Prevention of wheel vegetation destruction on certain depth becomes a new requirement of modern vehicles users, which has to be considered when procuring these vehicles. This means that negative influence of wheel tracks on vegetation should be blocked. This is very important for sensitive soil of forest areas with aim of avoiding the negative influence on ecological, economic and social function of vegetation. Soil deformation depth to 10 cm, protocol EcoWood suggests as environmentally acceptable. To evaluate environmental acceptability of different vehicles, as a relevant parameter general formula for wheel index (wheel numeric) is defined: $N_k = CI/p$.

Rut depth for one vehicle pass [3]:

$$z_1 = D\left(\frac{0.224}{N_k^{1.25}}\right) \tag{2.6}$$

Rut depth for multipass: $z_n = z_1 n^{1/a}$

D tyre diameter

z_1 first pass sinkage (m)

z_n sinkage after pass n (m)

n number of passes

p tyre inflation pressure

a coefficient of multipass (low soil load capacity $a = 2$–3; mean load capacity $a = 3$–4; high load capacity $a = 4$–5)

2.2.3.1 Vehicle Cone Index

Vehicle cone index (VCI) is accepted as referent NRMM model for unit mobility evaluation, developed by WES [2]. VCI is developed based on partially empiric term for ***MI*** *(Mobility Index)*. Experiments are used to determine the expressions for calculation of VCI values for one and for fifty indexes (VCI_1 and VCI_{50}). *VCI—Vehicle Cone Index represents minimum soil load (psi—pounds per square inch) within critical layer which enables that vehicle successfully completes certain number of passes.* In the US Army, expression *Soils Trafficability* is used, which means the capacity of soil to support military vehicles. Regarding mobility of one or more vehicles on the same track on typical low load terrain, soil trafficability is modelled, because vehicle mobility is related to humid coherent soil (fine grain soil/clay-loam, mud).

2.2.4 Mobility Index

$$MI = \left(\frac{K_{KP}K_T}{K_G K_L} + K_{OK} - K_{CL}\right) K_M K_{MJ} \tag{2.7}$$

K_{KP} contact pressure factor, $K_{KP} = w/0.00035\, n_1 d\, b$;
K_T vehicle axis load factor, $K_T = 0.073(w/1000) + 1.050$;
K_G tyre width factor, $K_G = (10 + b/25.4)/100$
K_L grouser factor *(without chain 1.00; with chain 1.05)*
K_{OK} wheel load factor, $K_{OK} = w/907.2$
K_{CL} chassis to soil distance factor, *H/254, H* distance between the ground and axle
K_M vehicle specific power factor *(>7.36 kW/t → 1.0; <7.36 kW/t → 1.05)*
K_{MJ} gearbox factor *(automatic = 1.0; manual = 1.05)*
w average axis load (kg)
n_1 number of wheels on the axis
b tyre width, non loaded wheel (mm)
d tyre diameter, non loaded wheel (mm).

Vehicle cone index

For one passage:

$$VCI_1 = 11.48 + 0.2MI - \left(\frac{39.2}{MI + 3.74}\right) 6.89655 \ \text{[kPa]} \tag{2.8}$$

For fifty passages:

$$VCI_{50} = 28.23 + 0.43MI - \left(\frac{92.67}{MI + 3.67}\right) 6.89655 \ \text{[kPa]} \tag{2.9}$$

Vehicle mobility and traction capability:

- for one vehicle pass: *CI > VCI*, "go"
- for multiple vehicle passes: *RCI = RI x CI > VCI*, "go".

Table 2.3 MMP and VCI for escort vehicles

Vehicle		Mass (kg)	MMP (kPa)	VCI_1 (kPa)
1	Land Rover Defender 110 TDi 4 × 4	3,000	343	172
2	MB G270 CDI 4 × 4	3,000	310	158
3	TAM 110 T7 4 × 4	7,000	358	207
4	TAM 150 T11 6 × 6	11,000	319	207
5	IVECO ML 100 E21 W 4 × 4	10,000	354	221
6	MAN 10.225 LAEC 4 × 4	10,800	368	234

(*Source* Ref. [4])

Table 2.4 MMP and NGP values for demining machines

Demining machine	Mass (kg)	MMP (kPa)	NGP (kPa)	CI_L (kPa)
MV-4	5,500	125	46	108
MV-10	18,000	150	39	124

Soil trafficability evaluation for some types of escort vehicles is given in Table 2.3, while in Table 2.4, value of *MMP* pressure, *NGP* pressure and CI_L index of light and medium demining machine, is given. A big difference is evident between MMP indicators of wheeled and tracked vehicles. Tracked demining machines offer significantly greater soft soil trafficability.

The basic demining machine *MV-4* has a nominal soil pressure of 46 kPa (≈0.46 bar) and mean maximum soil pressure 125 kPa (≈1.3 bar). Based on NGP value, quality assessment of soil trafficability is not possible. Comparing *MMP* machine value with the practically measured *CI* index, the use of the demining machine on certain terrain can be considered. Also, the required Limiting Cone Index (CI_L) can be evaluated, which determines the use of the machine: $CI_L = 0.827\ MMP$.

2.3 Toolbox Demining System

Demining machines using only a single tool are less useful than multipurpose machine which use more types of tools. Hence the development of demining machines is based on one basic machine of certain category and a greater number of different specialist tools for demining. Thereby a greater demining effectiveness is achieved. Such a universal working tool system and quick working tool replacement is called a toolbox demining system. Development of a demining working tool has established the following kinds of working tools, Fig. 2.2:

- Flail
- Tiller
- Combination of a flail and tiller
- Demining roller

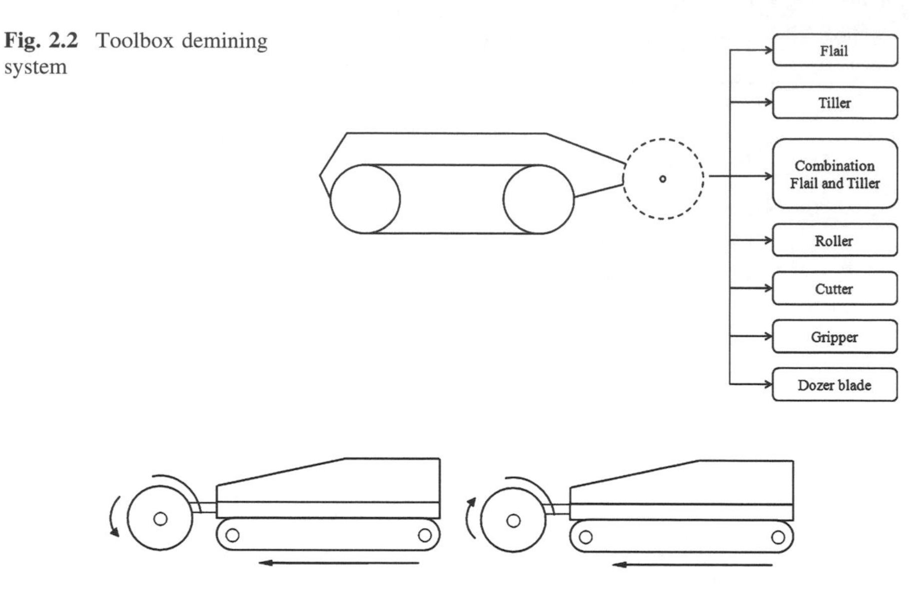

Fig. 2.2 Toolbox demining system

Fig. 2.3 Counter direction and same direction of rotor rotation to machine movement

- Vegetation cutter
- Gripper
- Dozer blade
- Other

Rotation of flail and tiller is in counter direction to machine movement, Fig. 2.3. Thereby during soil treatment a working tool impact is provided upon buried mines toward their neutralization. Rotation of working tool in the direction of the machine movement digs and throws out mines to the sides of the machine, offering a faster pass through the lane, to the expense of not actually destroying the mines. Such passage clearance through a minefield belongs to military demining rather than humanitarian demining.

A lower amount of engine power is needed for flail work than for the tiller. However, it is considered that tillers in certain categories of soil provide a greater quality in soil mine-clearance. Engine power has to be available to current machine movement resistances and operation of the working device. Hence, a demining machine of a specific category (light, medium, heavy) must have available power for usage of different tools in the toolbox system. Moreover, machines have to show a certain level of resistance to mine detonations. Light machines are to be resistant to AP mines, medium machines to AP mines and certain AT mines, whereas the heavy ones have to be resistant to all types of AP and AT mines.

The majority of demining machines use working tool switching, with flail or tiller, depending on the soil condition. A complex combination of working tools consisting of flail and tiller is used towards a greater reliability in the machine's neutralization of buried mines and its preservation from the impact of AT mine detonations.

A demining roller consists of several discs and they are used for opening up unsafe passages for vehicle convoys, and other. Other kinds of working tool are occasionally used for clearing of various obstacles in the demining process.

2.4 Soil Digging Resistance

Soil digging is based on sharp wedge, which has geometrical shape of digging blade. Basic parameters are: blade angle (20°–25°), back angle (5°–10°) and soil cutting angle (25°–35°). Lowest soil digging resistance provides semicircular shape and convex front part of blade. Soil digging process with working tools consists of two phases: soil cutting phase and soil disposition phase. In the soil cutting phase, soil layer is removed by the force of the tool blade. Cutting resistance is significant and depends on soil characteristics and tool blade condition. Soil disposition resistance on working tool depends on tool shape. Tool blade is a basic element of soil digging and mine destruction. With development of demining machines, different shapes of soil cutting/digging tools were also developed. For flails, hitting tool has a shape of a cutting and crushing hammer and for milling and tillers a shape of a wedge or a knife. During soil cutting phase, the soil is compacted first, than part of the layer shears in the plain of highest stress and after that soil is dispositioned.

2.4.1 *Soil Cutting Resistance*

When machine moves and digs the soil using working tools, (such as bulldozer, loader, and tractor), tool is subject to digging resistance. This resistance is much higher than resistance to machine movement so digging resistance requires deeper study. Digging resistance consists of cutting resistance (R_l) and displacement resistance of dug soil (R_2), different for each tool type. Cutting resistance is tangential component of overall digging resistance [4]. For supporting machine, bulldozers for example, cutting resistance can be formulated as:

$$R_1 = k_1 L h \ [\mathrm{N}] \tag{2.10}$$

k_l specific cutting resistance (N/m^2)
L cutting tool width (m)
h soil cutting depth (m)

Cutting resistance depends on the type of working tool (flail, tiller) and working conditions. Soil cutting shapes with flail and tiller are presented in Fig. 2.4. According to soil categorization, specific soil resistance (k_l) can range from 25 kN/m^2 for the first soil category up to 320 kN/m^2 in fourth soil category.

Fig. 2.4 Soil cutting: on the *left* tiller; on the *right* flail

Average values of specific resistance:

I. $k_l = 25$ kN/m^2 sandy clay, gravel
II. $k_l = 95$ kN/m^2 compact sandy clay, medium clay, soft coal
III. $k_l = 175$ kN/m^2 hard sandy clay with gravel, hard clay, conglomerate
IV. $k_l = 320$ kN/m^2 medium slate, hard dry clay, chalk and soft plaster stone, marl

To operate in hard soil category, teeth are mounted on toll blade, which loosen the soil and decrease cutting resistance for 25 %. It is important to properly set up digging depth and adequate distance between teeth in order to achieve less resistance.

Cutting resistance is dominant in relation to machine movement resistance (cca. 90 % of machine power is used for digging resistance). Cutting resistance increases more due to increase of cutting depth (h), then with increase of cutting width (L). In order to achieve required efficiency on certain soil category, machine operator should adjust cutting depth "h" and regularly inspect tool blades.

Soil cutting resistance [5]:

$$R_1 = k_1 b S_t \;\; [\mathrm{N}] \tag{2.11}$$

k_l specific soil cutting resistance (N/m^2)
b tool blade width, tooth
S_t current cutting layer thickness

2.4.2 *Flail Force Impulse*

Demining working device has a large number of flails attached to the rotor, rotating at 400–1000 rpm. Each flail has a striking hammer at the end of a chain for soil digging and mine neutralization. Hammer strikes on the soil can be viewed as analysis of collision of two bodies.

Based on *Law on Momentum Conservation*:

$$m_h v_o - m_h u = F_i \Delta t \tag{2.12}$$

Hammer behaviour during impact with soil can be described with collision factor k:

$$k = u/v_o \tag{2.13}$$

v_o circumferential hammer velocity before collision
u circumferential hammer velocity after collision

Factor k is related to soil categories, i.e. soil hardness, in order to notice the differences in digging in these soil categories (coefficient of restitution). Relative soil resistance according to soil hardness:

$$k_o = (1/k)100\,\% \tag{2.14}$$

Relation between the velocity before and after collision is provided through factor k, which provides the formula for calculation of hammer force impulse:

$$F_i = m_h v_o (1 - k)/\Delta t \ \ [\mathrm{N}] \tag{2.15}$$

F_i single hammer force impulse (N)
m_h hammer mass (kg)
Δt force impulse time interval of hammer soil grasping (s)

Hammer force impulse has to be higher than resistances for soil cutting or soil crushing. Hammer friction resistance causes decrease of hammer rotation speed which provides resistance factor k_O. Hammer rotation speed decreases due to friction resistance or inadequate power supply. When friction between hammer and soil is lost, hammer rotation speed increases. Increase of digging resistance is related to increased relative soil resistance. In these conditions, flail rotor requires more power.

2.5 Demining Machines with Flails

2.5.1 Flail Mechanics

Physical parameters of flails

General flail parameter is centrifugal force of striking hammer and chain. This force is applicable when choosing flail chain and rotor balance.

(a) *Flail centrifugal force*

$$\begin{aligned} F_{cf} &= m_h r\,\omega^2 + m_c r_s \omega^2 \\ F_c f &= 2\sigma d^2 \pi / 4 \end{aligned} \tag{2.16}$$

Chain diameter:

$$d^2 = 2\,\omega^2 (m_h r + m_c r_s)/\,\pi \tag{2.17}$$

m_h hammer mass (kg)
m_c chain mass (kg)
r hammer rotation radius (m)
ω angular hammer velocity (s^{-1})
d chain diameter
r_s chain mass radius ($r/2$)
σ strain of chain material

(b) *Hammer striking force*

$$\begin{aligned} F &= m_h a_t = m_h r \varepsilon \\ \varepsilon &= \Delta\omega / \Delta t \end{aligned} \tag{2.18}$$

$$F = m_h r \omega / t \tag{2.19}$$

a_t tangential hammer acceleration (s^{-2})
ε angular hammer acceleration (s^{-2})
ω angular hammer velocity (s^{-1})
r hammer rotation radius/flail radius (m)
t acceleration time (s)

Striking force F components, Fig. 2.5.
F_N normal component:

$$F_N = F\cos\theta \tag{2.20}$$

F_H horizontal component:

$$F_H = F\ \sin\theta \tag{2.21}$$

$\varphi + \theta = 90^0$
θ hammer striking angle (angle of chain)
Flail striking force acting on soil is tangential force F.

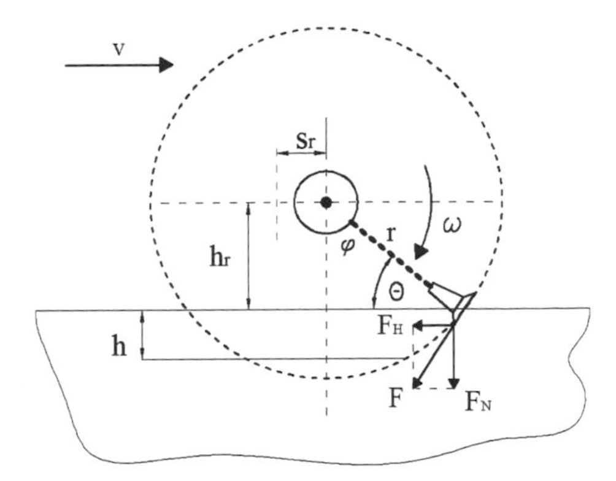

Fig. 2.5 Hammer striking force and parameters of flail

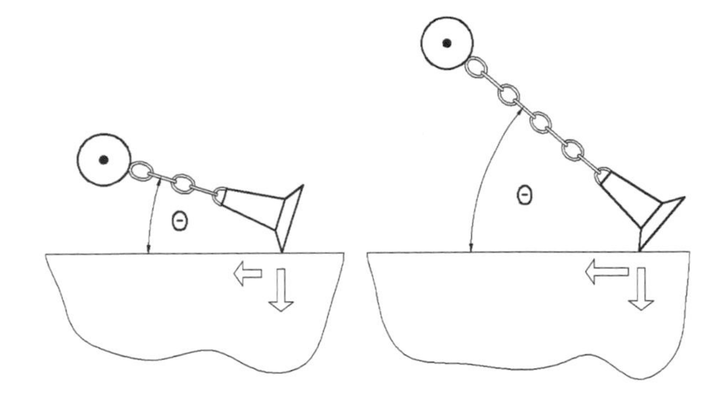

Fig. 2.6 Effect of hammer striking angle on the components of hammer striking force

Normal component of striking force $\boldsymbol{F_N} = F\ cos\theta$ depends on hammer striking angle θ. If striking angle θ decreases, vertical component of striking force increases. Accordingly, horizontal component of force $\boldsymbol{F_H} = F\ \sin\theta$ decreases, Fig. 2.6. The goal is to achieve required soil digging depth $\boldsymbol{h}$ with highest vertical striking force. That means that optimal height of rotor axis h_r can be determined as one of design parameters, based on normal component of striking force F_N.

Rotor axis height:

$$h_r = r \sin\theta \tag{2.22}$$

Relation between striking force and real conditions of soil treatment is practically expressed through hammer force impulse. Hammer force impulse has to be has to be higher than resistance at soil cutting or soil crushing.

Hammer impact to the soil can be analyzed through zone of soil stress below impact point. Shock waves are transmitted from impact point to under surface zone in expanding circles in all directions evenly, Fig. 2.7. Within this effect zone, mine can be detonated or crushed. On hard soil, shock waves are shallow, spreading more on the surface than into the deep ground. On medium hard and soft soil, the situation is opposite. Such a load distribution within the wave effect zone allows the possibility of mine destruction, using indirect contact of hammer—mine.

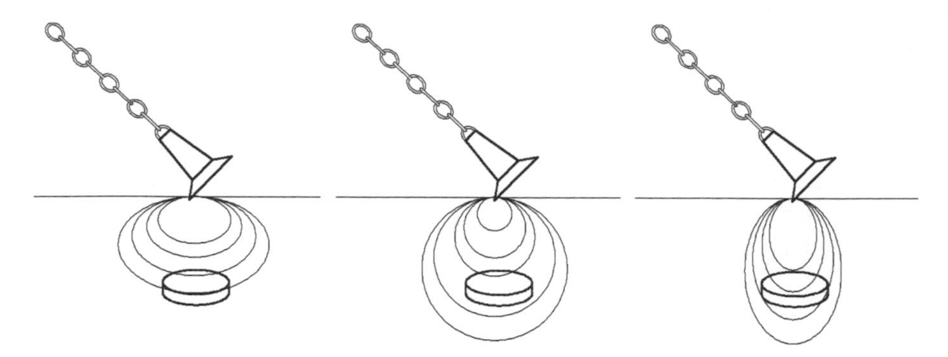

Fig. 2.7 Effects of soil on impact zones and soil stress distribution on a mine, hard soil, medium soft, soft soil

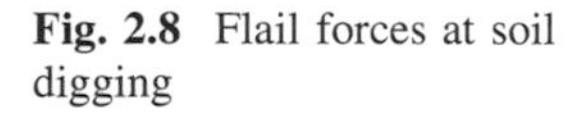

Fig. 2.8 Flail forces at soil digging

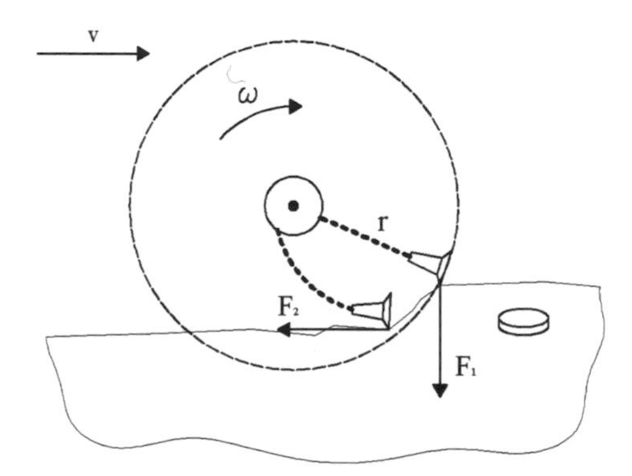

However, within interspaces of untreated area, where shock wave effect does not have influence, certain "pockets" can be found, so it is important to achieve necessary striking density in practice. Soft and medium hard soil require higher striking density. On hard soil, mines cannot be embedded deeper, and surface wave spreading may be adequate for mine destruction.

(c) *Empirical flail mechanics*

When analyzing flail impact on different soils, it can be found that such impacts are extremely complex and not thoroughly researched. Very important contribution in this area was provided by *Shankhla* research [6, 7].

Mechanics of flail is based on two main forces, Fig. 2.8:

F_1 —*Impact force,* hammer normal force when striking the soil. This hammer striking force provides penetration into the soil, which can neutralize embedded mines. Depending on tool type and tool blade geometry, force F_1 can cut certain soil or vegetation.

F_2 —*Drag force,* horizontal flail force which occurs after the force F_1; so called flail dragging force upon penetrated soil. This increases friction resistance between

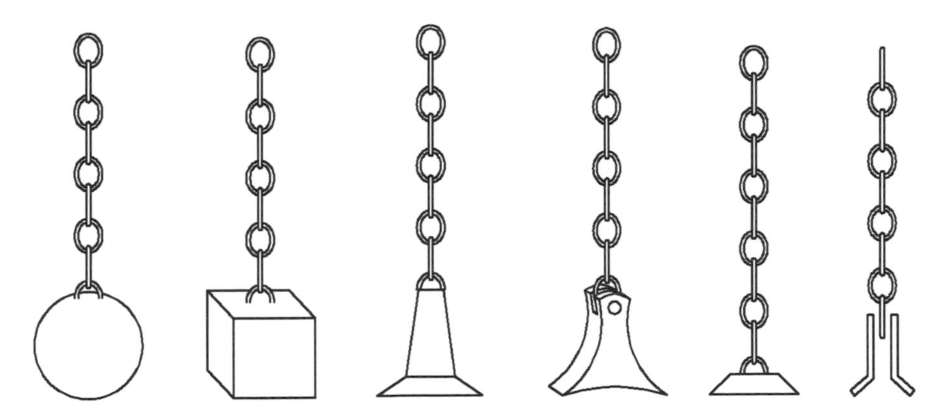

Fig. 2.9 Hammer shapes

flail and soil, which causes soil bulking and asymmetry in soil clearance. This force has negative influence on clearance performance; primarily, more power is needed for machine operation, and second, there is possibility of throwing the mines on already cleared area. Force F_2 causes problems to flail designers, because it cannot be exactly calculated. Magnitude of this force is based on assumptions, such as soil type, digging depth, shape and radius of working tool, rotation speed, etc. It is estimated that this force will double if soil is dug at depth of 20 cm instead of 10 cm. Additionally, extra weight is added approximately at chain mid length, in order to reduce the length of snake-shaped chain after the impact and reduce negative effect of dragging force. Smaller radius of flail provides lesser drag force, conversely a greater radius of flail increases the drag force, i.e. adds to the dragging force moment M_d. However, if there exists a zone of thick soil during treatment within a zone of soft soil, drag force will be greater than in treatment of homogenous soil.

2.5.2 Hammer Shapes

Geometry of striking hammer is highly important in the process of soil digging and mine neutralization. Common shapes of striking hammer are: *ball shape, block shape, mushroom shape, chisel shape and other shapes,* Fig. 2.9. Influence of hammer shape on demining process has not been thoroughly researched. Shape of striking hammer determines the position of striking force F_1 on soil and minimizing the force influence F_2, which causes drag resistance and soil shearing. Hammer shape is chosen according to soil category, digging depth, costs and experience. For example, ball shaped hammer may be ideal for areas where deeper penetration is not required and without vegetation. For deeper penetration, block shaped hammer could be very useful. However, when operating on medium soil category and for vegetation cutting, hammer shape is very important, for instance hammer shaped as a chisel. For soil cutting and mine neutralization, *mushroom shaped* hammer is favourable. There is no optimal hammer shape that could be used in all demining conditions. As mentioned earlier, mushroom shaped hammers are considered as universal. Optimal hammer shape is closely related to optimal soil digging depth, i.e. optimal penetration depth. Experience shows that there is no need to dig deeper than 10 cm, due to decreased efficiency in mine crushing, as well as negative effect of drag force. This also has significant influence on required power that has to be delivered to flail working tool, i.e. working costs.

(d) *Empirical flail mechanics*

Law on Momentum Conservation

$$m_h v_o - m_h u = F_i \Delta t \tag{2.23}$$

$$F_i = m_h v_o (1 - k) / \Delta t \tag{2.24}$$

F_i single hammer force impulse
F_{in} grasping hammers force impulse (refers to total impulse force the number of hammers that are engaged in the soil at any point in time)
m_h hammer mass
v_o circumferential hammer velocity $(r \pi n/30)$
k collision factor of soil category/ratio between hammer velocity before and after collision (factor of restitution)
Δt time interval of soil grasping hammers force impulse

Soil digging conditions

$$F_i > R_{\sigma i}, F_i > R_{ki} \tag{2.25}$$

Non-coherent soil resistance to crushing by hammer (approx.)

$$R_{\sigma i} = \sigma \cdot A \tag{2.26}$$

Coherent soil resistance to digging by hammer (approx.)

$$R_{ki} = z_n k_1 b \Sigma S_{ti} \tag{2.27}$$

z_n number of hammers in same digging position, two hammer helixes
k_1 specific resistance to soil digging, depending on soil category (N/m^2)
b hammer cutting width
ΣS_{ti} total thickness of soil digging layers $(S \sin\varphi)$
r flail hammer rotating radius
σ soil strength/normal stress (N/m^2)
$A = z_n b \Sigma S_{ti}$ surface of the hammer impact blade

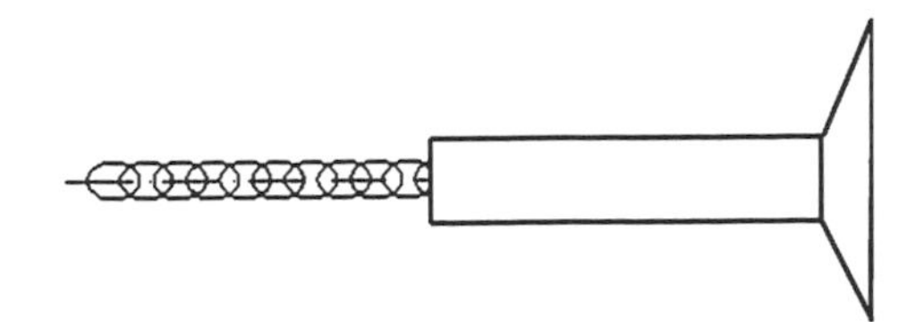

Fig. 2.10 Flail of demining machine (chain, hammer—"mushroom" shape)

2.5.2.1 Features of Hammer Force Impulse

Flail operation is based on impact force of the flail (flail strike) that strikes and digs the soil. Depending on soil humidity, soil is cut by hammer strikes and is moved towards machine shields. If the soil is dry and hard, soil is crushed and scattered. To simplify demining machine calculations, two theories of soil digging apply: theory of soil cutting and theory of soil hardness/crushing. Therefore, two tool types that are commonly used. For soft soil, cutting blade of lower weight is preferable, and for hard soil and cutting a rectangular shape of higher weight is advantage. Because of practicality, universal "mushroom" hammer shape is used. Soil that is coherent and soft can be cut using tool blade. Soil that is non-coherent, dry and hard can be crushed using different tool shapes. For the soft coherent soil, calculations from equation of soil cutting theory are applied, and for the hard non-coherent soil a criteria of soil hardness limit "σ" is applied. Hammer working principle is based on force impulse overcoming the resistance at soil digging (force impulse = change in momentum). In order to perform cutting, hammer force impulse has to be higher than resistance $F_i > R_{ki}$, i.e. for soil crushing condition of $F_i > R_{\sigma i}$ has to be fulfilled. Phase shift between flails (attached to rotor's helix/spiral (n × 180°) starting from the rotor centre towards the rotor edges) decreases digging resistance, removes the influence of axial forces and unbalance of the flail. These flails are used on demining machines.

Total resistance momentum to flail rotation includes static and dynamic momentum of flail's parts rotation, until hammer strikes the soil, when kinetic energy is lost, angular velocity is changed and flail rpm is decreased. Since this change in energy is not well known in practice, each flail cyclical operation is assumed: hammer acceleration and stopping at cutting the soil layer.

The striking hammer is of "mushroom" shape with a cutting blade, Fig. 2.10.

Hammer Impact to the Soil

Hammer strikes to the soil can be viewed through two analyses of two body collision. It is necessary to determine the hammer force impulse. One method of determining this force assumes that force is acting in finite time interval, in which hammer is using part of its momentum it had before impact. It can be assumed that hammer behaviour during impact with soil can be described with collision factor:

$$k = \frac{u}{v_0} \tag{2.28}$$

v_o circumferential hammer velocity before impact
u circumferential hammer velocity after impact

Through the introduction of collision factor k problem of hammer striking the obstacle can be viewed as a problem from classical mechanics, not using the assumption of a firm obstacle. Factor "k" is within interval:

$$0 \leq k \leq 1 \tag{2.29}$$

$k = 0$ impact is ideally plastic
$k = 1$ impact is ideally elastic
$k_o = (1/k)$ 100 % relative soil resistance

Based on Eq. (2.23), ratio between velocity before and after collision is provided through factor k, leading to equation for force impulse in time interval, Δt (2.24): $F_i = m_h\, v_o\, (1\ k)/\Delta t$.

Hammer Force Impulse Analysis

Example: digging depth $h_1 = 100$ mm, hammer weight, $m = 1.2$ kg, $\varphi = 35°$, $n = 900$ rpm.

Assumption:

- factor k for 3 different collision conditions; $k = 0.3$; 0.5; 0.7; relative soil resistance; $k_o = 333$; 200; 143
- time interval of force impulse impact in interval up to $\Delta t = 10^{-4}$ s

Using factor k different situations for the working tool could be described, from striking the obstacle (i.e. how much energy is lost because of that), to the situations when hammer strikes the mine and the factor k sign changes. Accordingly, depending on the soil type, obstacle k, digging depth h and flail rotation n, calculations could be performed of force impulse influencing the hammer when striking the soil,.

Factor k can be brought into relation to the soil category, i.e. soil hardness, in order to determine the differences between them. Coefficient k can be simulated as well as digging depth h, under the assumption that $k = 0.7$ for III soil category, $k = 0.5$ for IV soil category and $k = 0.3$ for V soil category (special). Hammer force impulse of one hammer F_i for specific digging depth is multiplied with number of hammers that are grasping the soil. Increasing the hammer's rpm for the same factor k, force impulse $F_i = f(n)$, $\Delta t = 1 \times 10^{-4}$ s, increase is linear, Fig. 2.11.

For hammer force impulse in the time interval $F_i = f(\Delta t)$ it can be concluded that, when hammer strikes the soil, force impulse decreases. The force is highest when hammer is striking the soil, and is decreasing during removal of the dug soil layer. It can be assumed that with decrease of coefficient k on the flail's rotor shaft additional power will be required for acceleration of the lagged flail. Theoretically, hammer rotation speed can decrease to zero value. It is not possible to quickly reestablish hammer speed due to dragging force that appears at hammer's reacceleration. Practice shows that such hammer lagging causes rapid wear-out and chain

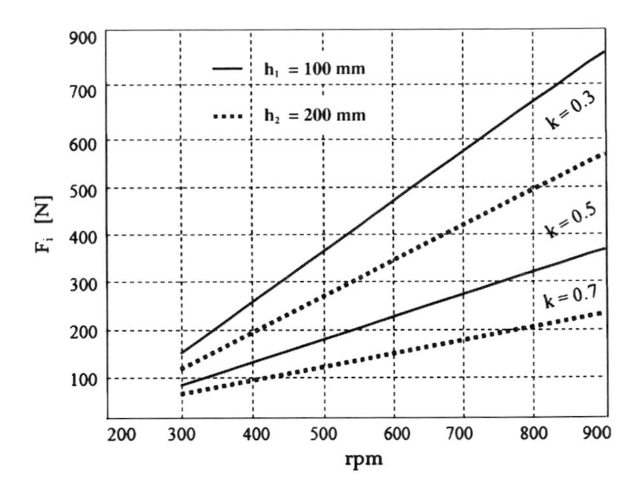

Fig. 2.11 Hammer force impulse in relation to k, as a function of rpm $F_i = f(n, t)$

Table 2.5 Grasping hammers force impulse at soil digging F_{in} [N]

Digging depth h [mm] n rpm	k	$h_1 = 100$ mm ($z = 6$) $\varphi_1 = 35°$	$h_2 = 200$ mm ($z = 8$) $\varphi_2 = 50°$
$n = 200$	0.3	1,009.75	1,003.50
	0.5	721.25	716.79
	0.7	432.75	430.07
$n = 500$	0.3	2,524.38	2,508.75
	0.5	1,803.13	1,791.96
	0.7	1,081.88	1,075.18
$n = 900$	0.3	4,543.88	4,515.75
	0.5	3,245.63	3,225.53
	0.7	1,947.38	1,935.32

elongation (in practice—when hammers are replaced because of wear-out, chain elongation is 10 %). Finally, with relative restitution coefficient k_O increment, rotor's resistance increase, meaning that flail rotor requires more power. Grasping hammers force impulse is given in the Table 2.5. It refers to the total impulse force of the number of hammers that are engaged in the soil at any point in time.

2.5.3 Machine Power

Power required for flail rotor rotation

$$P_r = M_u \omega \, [\mathrm{W}] \tag{2.30}$$

M_u— total flail rotation resistance moment, includes:

$$M_u = M_{st} + M_{din} + M_d \text{ [Nm]} \quad (2.31)$$

$M_{st} = M_e + M_\mu + M_g; M_{din} = M_i$
M_e = *moment of eccentric rotor mass*
$M_e = \Sigma_{i\text{-}n}\, m_{ri}\, g\, r_{ri}$
M_μ = *friction moment in shaft bearing (insignificant)*
$M_\mu = F_\mu\, r_v = \Sigma_{i\text{-}n}\, F_{ci}\, \mu\, r_v$
M_i = *moment of shaft, hammer and chain inertia*
$M_i = J\omega/t + \Sigma_{i\text{-}n}\, m_i\, a\, r_i = \Sigma_{i\text{-}n}\, m_i\, a\, r_i + J\varepsilon$
M_d *moment of flail dragging force* (F_2)

$$M_u = \Sigma_{i-n} m_{ri} g\, r_{ri} + \Sigma_{i-n} F_{ci} f_t r_o + \Sigma_{i-n} m_i a\, r_{oi} + J\varepsilon + M_d \text{ [Nm]}$$

m_i striking hammers and chains mass (kg)
r_i flail hammer rotating radius (m)
J shaft total inertia moment (rotor, hammers, chains) (kgm^2)
ω angular rotor velocity (s^{-1})
ε angular acceleration (s^{-2})
F_{ci}, f_t centrifugal hammer force, resistance factor
r_i (r) rotating flail radius

Power required for machine movement

$$P_v = \Sigma_{i-n} R_i v \text{ [W]} \quad (2.32)$$

$\Sigma_{i\text{-}n}\, R = R_k + R_i + R_\alpha$
$R_k = G\, cos\alpha\, f_k$, rolling resistance (wheels/tracks)
$R_i = m\, a$, inertia resistance
$R_\alpha = G$ sinα, slope resistance
v = machine speed

Total power required for machine movement and operation:

$$P_T = P_r + P_v \text{ [W]} \quad (2.33)$$

Most of the machine power is used for soil treatment using flail, and less power for machine movement. *Engine power* of machine is increased due to losses in power train and additional devices (η_p):

$$P_m = \frac{P_T}{\eta_p} \text{ [W]} \quad (2.34)$$

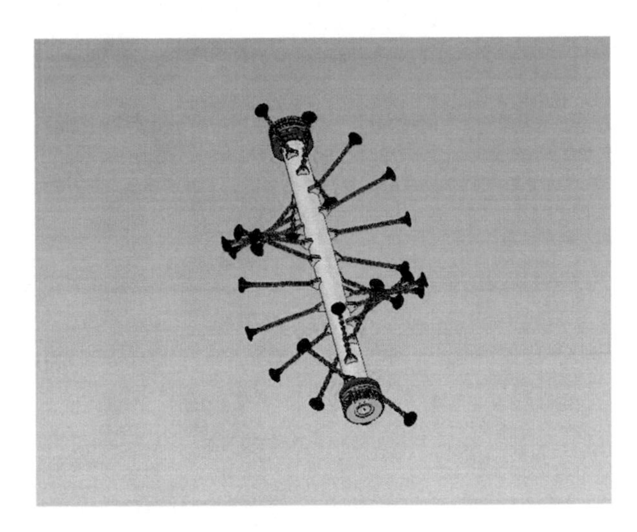

Fig. 2.12 Flail demining helix system

2.5.4 Flail Design

The best soil digging density is achieved by hammer shear of 16 mm, where machine movement speed is less than 1 km/h. Higher machine speed is achieved by hammer shear of 30 mm, which provides for better working efficiency. Machine should maintain required soil digging density (depth, 10, 15, 20 cm), with possibility to lower its movement speed if flail rotor rpm suddenly decreases.

On flail rotor of 2.5 m digging width, 50–70 flails could be attached on several helixes with calculated positioning of flails, which will provide required soil digging density, Fig. 2.12. Flail strikes overlaying for few millimetres ensure that each mine will be destroyed for the whole flail width, i.e. even smallest mine or explosive ordnance will not be missed.

Good machine speed is 0.7–0.8 km/h, and flail rpm is 600–800. In theory, resulting untreated area is extremely small in relation to the size of the smallest mine or its fuse, meaning that the risk of mines not destroyed is very low. Tool shear remains constant; increase in machine movement speed is resulting in increase of rotor rpm, and vice versa.

Most common are machines with flails for soil digging to a certain depth, during which buried mines can be neutralized. In practice, detailed knowledge and understanding of flail design and operation is still not sufficient. That is why evaluation of flail performance, i.e. digging depth, soil treatment quality and ability to destroy smallest AP mines, is done through testing. That provides an analytical model for flail calculations and design of basic flail parameters. Technological working speed and flail rpm are adjusted, interaction forces between flail hammers and soil based on force impulse of flail hammer, and shape of flail hammer and chain in relation to their durability are defined, and dimensions of working tool rotor and flails are determined. Calculation results need to be verified through testing of light and medium demining machines. The flail system has to destroy embedded AP and AT mines. Effects of AP mine explosion

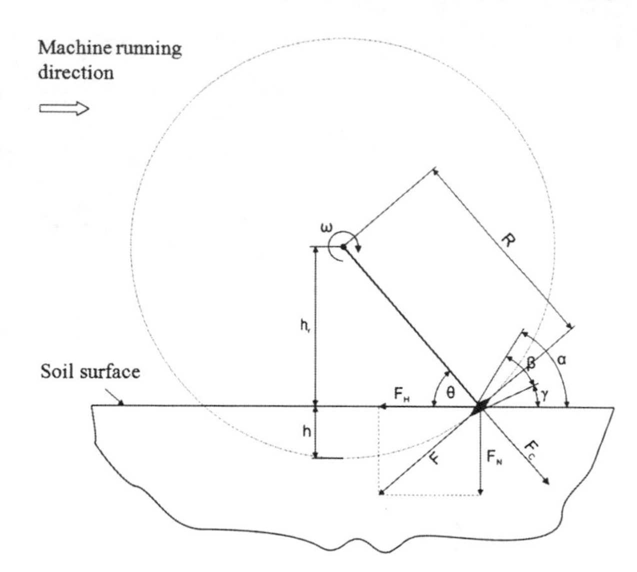

Fig. 2.13 Striking hammer force and geometric parameters of flail; θ hammer striking angle, α cutting angle, β wedge angle, γ back angle, h soil digging depth

shouldn't damage the flail. AT mine activation can cause damage to some chains of the flail. Rotor of the flail system has to remain undamaged and the machine is ready for further demining.

Flails on rotor are mounted along the helix path on the rotor shaft. Helix starts from the middle of the rotor towards the end on each side. Often, two helices with phase shift of 180° are set up. Basic flail dimensions are: flail diameter D, rotor width b, number of helix n_Z, and number of flails/hammers z. In order to reduce number of grasping hammers, number of helix can be increased. Good flail positioning reduces number of grasping hammers and soil treatment resistance.

2.5.4.1 Flail and Soil Interaction

Striking hammer force and geometric parameters of flail are presented in Fig. 2.13. Tangential force F causes hammer penetration and compressive soil disturbance. According to second Newton's Law, follows:

$$F = m_h a_t \text{ [N]} \tag{2.35}$$

m_h hammer mass,
a_t tangential hammer acceleration, $a_t = r\varepsilon$
r hammer CG rotation radius around rotor's axis,
ε hammer angular acceleration

Flail's task is to treat the soil with certain digging depth and neutralization of AP and AT mines. Normal component of force F_N should be as high as possible in

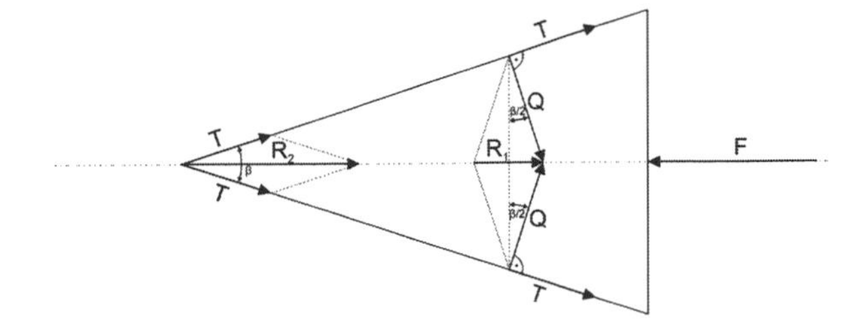

Fig. 2.14 Forces on the wedge when digging the soil

order to dig the soil and crush embedded mines. Force F_H is function of angle sinus θ, and can be decreased, but not too much, because without this force hammer penetration into the soil could not be achieved.

Hammer Force Impulse According to Law on Momentum Conservation

Relation between circumferential hammer velocity before and after collision can be established through soil resistance factor ($k = u/v_o$; v_o—circumferential hammer velocity before collision, u—circumferential hammer speed after collision). Collision factor may be perceived as soil resistance factor, because it can be brought into relation with soil hardness, in order to distinguish the differences for digging of certain soil categories.

Hammer impulse force:

$$F_i = m_h v_0 (1 - k)/\Delta t\,[N],\ \Delta t - \text{impulse duration time}$$

Hammer impulse force has to be higher than cutting resistance or soil cutting resistance, i.e. soil digging condition has to be fulfilled → $F_i > R_{ki}$, i.e. $F_i > R_{\sigma i}$. (R_{ki}—soil cutting resistance/coherent soil, $R_{\sigma i}$—soil crushing resistance/non-coherent soil).

2.5.4.2 Flail Modelling

A part of hammer that strikes the soil can be modelled in a shape of wedge, Fig. 2.14. On the wedge surface, stress pressure force Q appears as reaction to hammer striking force—hammer impulse force. These forces cause digging resistance.

Resultant force R_1 is opposed to tangential hammer force F and represents a part of digging resistance. Stress forces Q acts vertically on wedge surface and cause friction forces T, which appear on the wedge surface. Their resultant is force R_2 which represents a part of digging resistance. The wedge efficiency is determined with ratio of wedge force without friction R_1 and required wedge force F [8]:

$$\eta_w = \frac{R_1}{F}$$

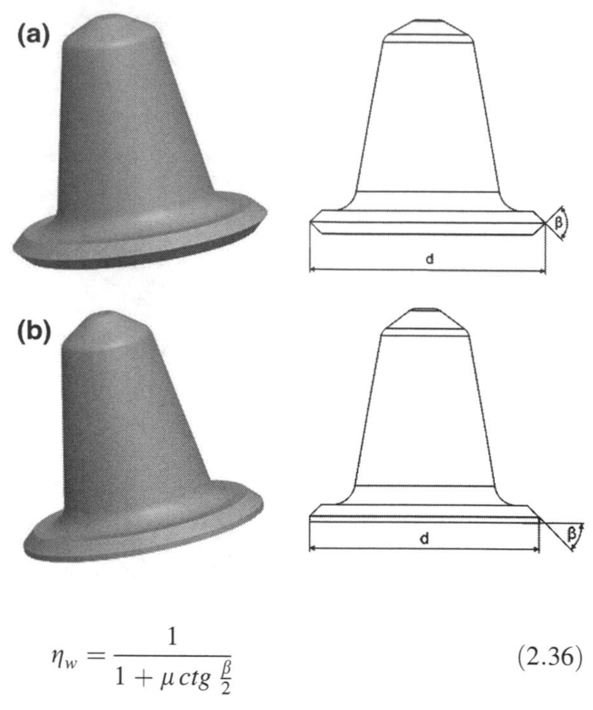

Fig. 2.15 Optimal shapes of striking hammer, hammer diameter d and wedge angle β

$$\eta_w = \frac{1}{1 + \mu\, ctg\, \frac{\beta}{2}} \tag{2.36}$$

The wedge efficiency is higher if friction coefficient μ is lower between the wedge and the soil, and if angle β is higher. For humid plastic soil, smallest digging resistance can be achieved at smaller angles, and for hard soil at higher cutting angles. This means that in this area the wedge efficiency can be observed. It can be assumed that optimal shape of striking hammer for soil treatment in all soil categories is shape of a bell or mushroom, Fig. 2.15. Bell blade vertical cross section has a shape of wedge, and horizontal has cross section shape of a circle. This causes the smallest soil cutting resistances. Hammer's centre of gravity (CG) is placed on hammer axis approximately at *(1/5-1/3)L* from the hammer base. Hammer head diameter (*d*) is 50–60 (90) mm. Favourable hammer shape for lighter soil categories is shown at Fig. 2.15b. Blade of such hammer has a shape of a wedge, designed so that lower part is vertical to strained chain. This wedge shape enables that necessary cutting angle α can be achieved with smaller flail radius r.

In practice, working tool shapes are simple, in order to manufacture them more easily and of less cost, Fig. 2.16. Material, from which hammers are made, is usually steel for cementing, EN 16MnCr5. With cementing, hard surface layer is achieved, resistant to wear out, and core retains its ferrite—perlite structure, which is tough and resistant to dynamic and strike loads. Striking hammer exploitation life cycle is 40,000–50,000 m^2 of treated soil, i.e. around 50 working hours, after which they should be replaced. Hammer mass is between 0.75–1.5 kg, depending on soil hardness. Flexible connection between shaft and hammer is a chain of 12–15 mm in diameter. Chain's exploitation life cycle is around 80,000 m^2 of treated

Fig. 2.16 Working shapes of striking hammers (*Source* CROMAC-CTDT)

soil, i.e. around 80 working hours, after which it should be replaced. At the contact surface of the chain links, chain extreme wearing out is present, which causes its elongation during operations. Chains that are used for flails are usually made of steel used for improvement EN C45, and is heat treated and tempered and in order to achieve high tensile and yield strength, while retaining toughness and dynamic durability.

2.5.5 Flail Geometry

When determining optimal rotor and flail dimension, the goal is to achieve required soil digging depth h with highest standard force F_H. From this fact, it can be concluded that based on standard striking force F_N component, an optimal rotor axis height h can be determined as a design parameter.

Rotor's height from the ground, Fig. 2.13, is:

$$\begin{aligned} h_r &= r \sin\theta [\text{m}] \\ h_r &= (h + h_r)\sin\theta \end{aligned} \tag{2.37}$$

Hammer impact angle:

$$\sin\theta = \frac{h_r}{h + h_r}$$

$$\theta = arc\sin\frac{h_r}{h + h_r} [^\circ] \tag{2.38}$$

Soil digging depth h is usually known and is 10–20 cm. When calculating dependence of certain flail parameters, soil digging depth h = 10–20 cm and rotor radius r = 400–1000 mm were simulated. Diagram of flail parameters is shown on the Fig. 2.17.

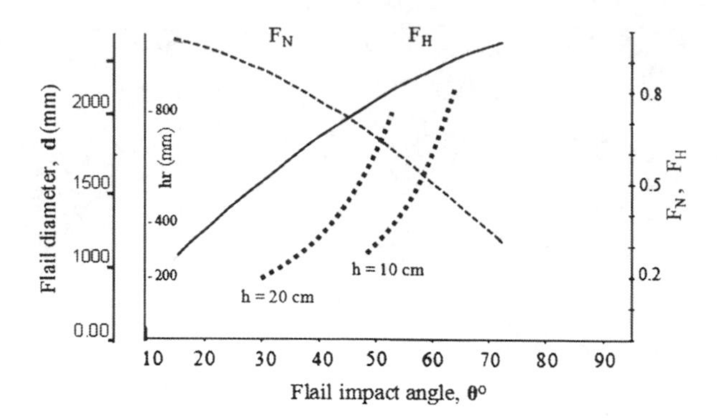

Fig. 2.17 Flail parameters in function of hammer impact angle θ

If rotor height h_r increases, flail angle at which hammer strikes the soil θ increases too. The wedge angle β is chosen according to soil hardness:

- angle of soil cutting: $\alpha = \beta + \gamma$, $\beta = \alpha - \gamma$
- back angle: $\gamma = 90° - \beta/2 - \theta$

When digging resistance increases, rotor height h_r should be increased too, which further increases striking force for the same hammer angular speed. Dependence between hammer striking angle θ and soil cutting angle α can be considered. This dependence determines if the soil is treated by cutting or crushing. For digging soil of lighter category a greater cutting angle α needs to be set, i.e. a lesser hammer impact angle θ, and conversely. For determined wedge angle (e.g. $\beta = 20$–$40°$) dependence between impact angle θ and cutting angle α is linear, e.g.:

- for wedge angle $\beta = 20°$, hammer impact angle is $\theta = 70°$, and cutting angle is $\alpha = 35°$
- for wedge angle $\beta = 30°$, hammer impact angle is $\theta = 60°$, and cutting angle is $\alpha = 45°$

2.5.5.1 Rotor Width and Quantity of Helixes and Flails

Rotor should be designed in a way that number of hammers in grasping operation is the minimum number of hammers, because digging resistance is the lowest, and required engine power is lower, which results in lower fuel consumption and higher efficiency. Number of grasping hammers depends on rotor radius r, and digging depth h. Axial hammer distance is determined by soil treatment density requirements and is usually $l_u = 0$–15 mm. If hammer distance is equal to zero, than there is no difference between the strikes in transverse direction, and if distance is e.g. 15 mm, than distance between strikes in transversal direction is 15 mm. This is acceptable, because the length of mine fuse that is to be crushed is 16 mm, and radius of the smallest AP mine (PMA-2) is 68 mm. Increase of axial

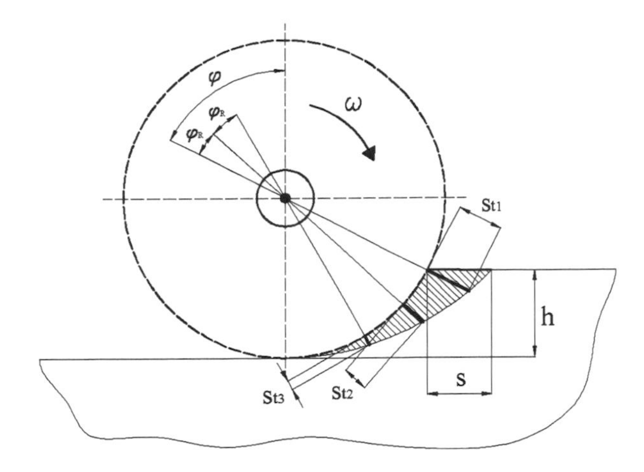

Fig. 2.18 Hammer digging of coherent soil/hammer's phase shift

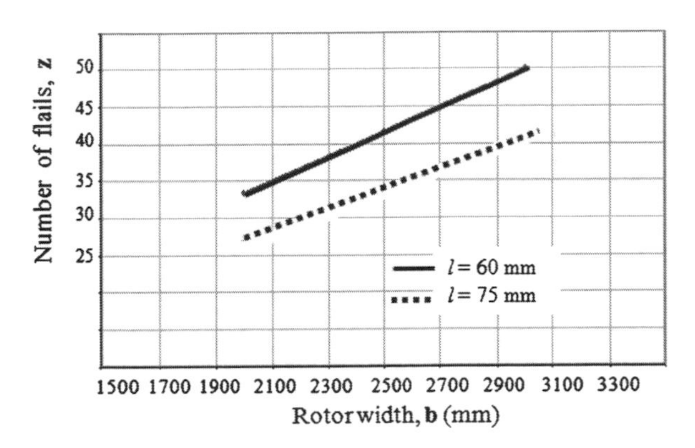

Fig. 2.19 Number of flails/hammers z in function of rotor width b and axial flail distance l on the rotor

hammer distance causes decrease of number of grasping hammers, as well as digging resistance.

Calculated values of hammer blades in grasp φ, hammer phase shifts φ_R, number of grasping hammers z_m and required quantity of flails/hammers z on one helix for different rotor width $b = 2000\ldots3000$ mm, number of helixes on rotor $n_z = 1\ldots3$, axial flail distance $l = 60\ldots75$, axial striking hammer distance $l_u = 0\ldots15$ mm, hammer diameter $d = 60$ mm and digging depth h = 100…200 mm, were simulated too (Figs. 2.18, 2.19).

Angle of hammer chain

$$cos\varphi = (r - h) / r \tag{2.39}$$

Hammer's phase shift/angle between two hammers

$$\varphi_R = 360l / Z \quad (2.40)$$
$$Z - helix\ step$$

Number of helixes on rotor

$$n_z = 1 \ldots n_n \tag{2.41}$$

Number of flails on one helix

$$z = 360/\varphi_R \tag{2.42}$$

Number of hammers in grasp

$$z_m = \varphi/\varphi_R \tag{2.43}$$

Axial flail distance between hammers

$$l = (2d + l_u)/2 \tag{2.44}$$

d hammer diameter,
l_u distance between hammer grasps

Machine working speed and rotor rpm

$$v = Sn, n = v/S \tag{2.45}$$

S hammer shear,
v machine speed, n rotor rpm

Current thickness of dug soil layers

$$\begin{aligned} S_{t1} &= S \sin \varphi \\ S_{t2} &= S \sin(\varphi - \varphi_R) \\ S_{t3} &= S \sin(\varphi - 2\varphi_R) \end{aligned} \tag{2.46}$$

If rotor width increases, total number of rotor flails increases too. If axial flail distance on rotor is higher, number of flails decreases. Rotor width is important parameter from the machine efficiency point of view, because machine with wider rotor can treat the soil faster. Rotor width is constant, and digging depth depends on users' requirements, flail optimization is done by selecting the number of helixes and rotor radius. At the end of analysis and flail dimensions calculation, *optimal flail parameters* can be estimated.

For soil digging depth of 20 cm and flail width up to 2 m, flail radius is between 0.75–2.0 m. According to this parameter, necessary flail striking force or impulse force, which enables soil cutting, can be determined. Number of striking hammers is 25–40 for the rotor width of up to 2 m, and they can be placed on two or more helices. Striking hammer mass, regarding its volume, is 0.6–1.5 kg. Based on analytical calculation and machine design, it can be concluded that considered flail calculation model is adequate. Relevant values of flail parameters regarding soil digging criteria and AT mine detonations are provided in Table 2.6.

Table 2.6 Relevant values for flail design parameters

Rotor width, b	2.0 m	2.5 m	3.0 m
Rotor radius, r	400–1000 mm	400–1000 mm	400–1000 mm
Digging depth, h	to 200 mm	to 200 mm	to 200 mm
Number of hammers, z	33 (27)	42 (33)	50 (40)
Hammer weight, m_h	0.6–1.5 kg	0.6–1.5 kg	0.6–1.5 kg
Hammer diameter, d	60 mm	60 mm	60 mm
Number of helix, n_z	1–3	1–3	1–3
Distance between strikes, l_u	0 mm (15 mm)	0 mm (15 mm)	0 mm (15 mm)
Rotor rpm, n	300–1000 min^{-1}	300–1000 min^{-1}	300–1000 min^{-1}
Working speed, v	0.5–1.7 km/h	0.5–1.7 km/h	0.5–1.7 km/h

Table 2.7 Parameters of the MV-4 flail system

Light machine	MV-4
Rotor width, b	1800 mm (2015)
Rotor radius, r	450 mm
Digging depth, h	200 mm
Number of hammers, z	27 (34)
Hammer shape and weight, m_h	Bell-shaped hammer, 0.8 kg
Hammer diameter, d	95 mm (60 mm)
Number of helixes, n_z	3 (2)
Distance between hammer strikes, l_u	48 mm (34.5 mm)
Rotor rpm, n	0–900 min^{-1}
Working speed, v	0.5–2 km/h

2.5.5.2 Mine Neutralization and Flail Durability

Calculation results were verified on development and testing of the MV-4 demining machine. Parameters for soil treatment of the working unit MV-4 are provided in Table 2.7. Neutralization of AP mines after soil treatment is shown on Fig. 2.20.

Effects from AP mine explosion are not significant and do not damage the flail. However, AT mine activation, e.g. of 6 kg TNT can cause major damage to the flail system. For low rotor position above the ground (small r) damage to the working device is major, requiring replacement of whole device. Evaluation of "cone of destruction" of AT mine effects has been done, according to NATO II level protection (6 kg TNT). Evaluation is based on *MV-4* demining machine testing results.

To verify flail durability, reflective blast pressure p_{ur} which appears under the flail, is applicable [9]. To verify striking hammer durability, pressure of detonation products p_d is applicable, and to verify flail sides, a pressure of a blast wave p_{us}, is applicable, Fig. 2.21. At the distance of rotor shaft from the ground of 50 cm, reflective blast pressure p_{ur} of AT mine with 6 kg TNT-a, is up to 60 MPa (600 bar). Regarding tests performed, this pressure does not have important influence on the rotor's shaft of 150 mm in diameter, except for the few damaged flails which were in direct contact with the mine. However, if explosion happens at

Fig. 2.20 Neutralized mines, crushed mines (Reproduced with permission from CROMAC-CTDT)

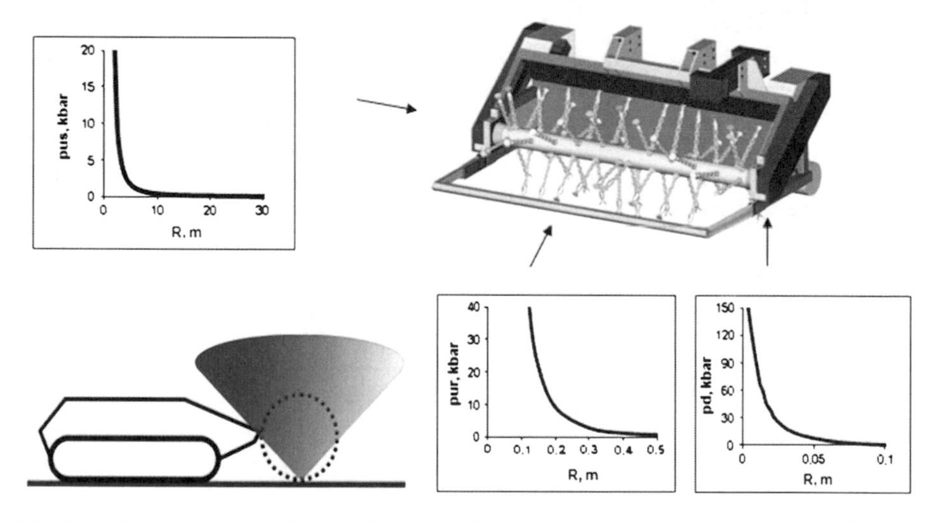

Fig. 2.21 Blast test—AT mine striking wave effect on flail, under the rotor, on the end of the rotor, and aside the rotor

the end of the shaft, there is possible damage to the shaft bearing. When testing MV-e flail durability against a more destructive mine (8 kg of TNT) under the rotor, 4–5 flails were damaged. Rotor itself had been undamaged and the machine was ready for further operations [10].

2.5.5.3 Conclusion

1. Design of flail system is done based on soil category that will be treated using machines and requirements for digging depth h and resistance to detonation of AT mines. If treated soil is of lighter category, than rotor radius can be smaller. If treated soil is of heavier category, than rotor radius has to be bigger in order to provide adequate striking force F to overcome digging resistance.
2. Wedge angle β is selected according to soil category that has to be treated, and for lighter soils that are treated by cutting, a smaller angle β is selected, while for harder soils that are treated by crushing, a higher wedge angle β is selected. Additionally, for treatment of lighter soils, hammer of light weight are used, and for soils of harder categories hammers of heavier weight are used.
3. Rotor width depends on user's requirements for machine working efficiency. If rotor is wider, working efficiency U is higher. Number of flails with hammers for soil digging is $z = 25–30$ for digging width of $b = 2$ m. Soil treatment density depends on machine movement speed v and rotor rpm n, i.e. on longitudinal distance of striking hammers l_u and distance between the hammer strikes. Optimal flail diameter is within $D = 1–2$ m for digging of all soil categories at depth down to $h = 30$ cm.
4. "Cone of destruction" of AT mine with 6 kg of TNT destroy several flails on the rotor, but does not damage the rotor. Explosion at the end of the flail can cause damage that could require rotor replacement.

2.6 Demining Machine with Tillers

Tillers design is based on principle of digging the soil and mine neutralizing. Soil digging mode shows rotor relative rotation in relation to machine movement, i.e. shows direction in which the soil under the rotor is thrown out, Fig. 2.22. Counter-direction mode of soil digging is used for neutralizing mines under the rotor, and the same direction mode is used for neutralizing the mines in front of the rotor, Fig. 2.23. When digging the soil in the same direction mode cutting force is trying to lift the object off the ground, and mine can be crushed, detonated or ejected. In counter-direction soil digging mode, tiller tries to press the object into the ground, and mine can be crushed, detonated or embedded deeper. Regarding tool durability, digging in the same direction is better for the soil with hard surface layer, because tool blades are first digging through the softer soil under the hard surface layer. In counter-direction mode, tool blades are striking the hard surface layer first.

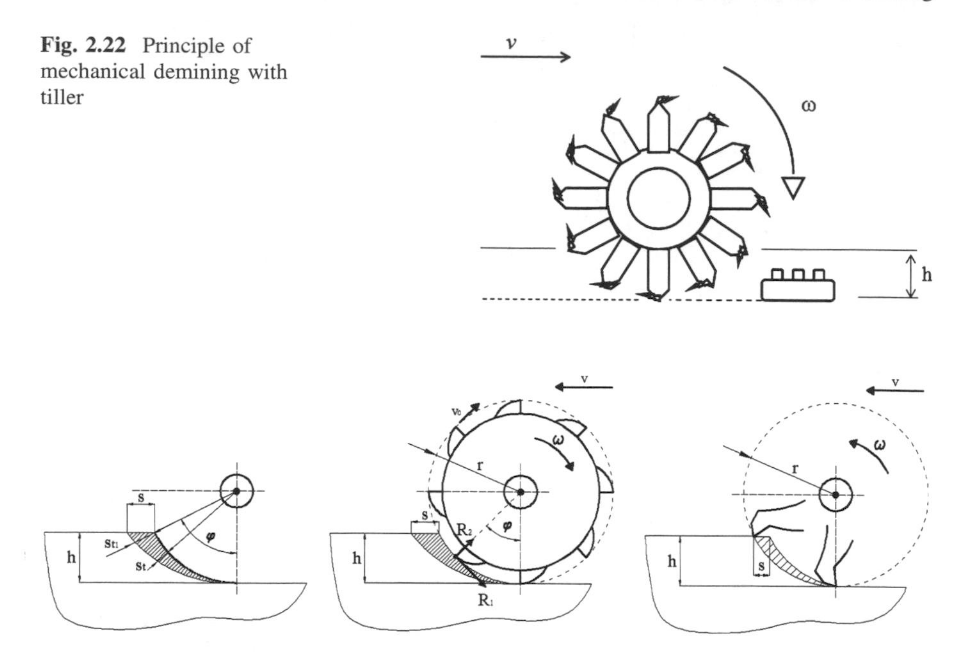

Fig. 2.22 Principle of mechanical demining with tiller

Fig. 2.23 Soil digging—in the same direction and counter direction, shape of cut layer

2.6.1 Helix System

Positioning model of tiller tool has a shape of helix bolted along the rotor (to the left and to the right from the rotor centre) with sharp teeth for soil digging and mine neutralizing. Helix acclivity angle is $\alpha = h/2\ r\ \pi$, where h—helix pitch, and r—radius of tiller rotor. Along this helix, teeth are attached to the rotor. Density of mounted teeth and tiller cutting resistance depends on helix acclivity. Besides milling, one part of unmilled soil has uniform movement along the rotor, due the helix, acting as tape-loop. For each rotor revolution, material is moved forward for the distance equal to helix pitch.

Teeth blades should be positioned tangentially in relation to tiller rotor. Positioning of teeth blades on rotor depends on object that should be destroyed. For AP mines, critical size or tool shear of 16–30 mm is assumed. Size of 16 mm is strictly set up, for destruction of smallest mine fuses. Distance between teeth blades on circumference rotor lines has to be less then critical value. From the mine-clearance diagram technological machine movement speed and rotor rotation speed are determined.

2.6.2 Cutting Resistance

When digging, most of the teeth blades are grasping the soil. Number of grasping teeth blades depend on digging depth. Uneven number of tools that are grasping the soil, and change in depth of soil cutting, is causing asymmetric digging resistance and asymmetry in required rotor torque. Soil digging consists of two phases, *cutting phase* and soil layer *displacement phase*. Total digging resistance of the first phase consists of tangential component in direction of tangent on blade trajectory, and perpendicular vertical resistance component directed towards rotor axis, i.e. radial components. Displacement resistance between material and teeth along the rotor helix is not negligible, because part of excavated material is moved along the tool.

Cutting resistance is tangential component of digging resistance

$$R_1 = k_1 b\, S_t \ [\mathrm{N}] \tag{2.47}$$

k_l specific cutting resistance
b tool blade width, teeth
S_t cutting layer thickness

Total cutting resistance m—blade in grasp

$$R_{1u} = \sum\nolimits_{1-m} k_1 b\, S_{ti} \ [\mathrm{N}] \tag{2.48}$$

Normal force is radial component

$$R_2 = (0.2 - 0.6) R_1 \ [\mathrm{N}] \tag{2.49}$$

Specific cutting resistance depends on tool shape and condition of its blade. More significant for resistance is increase of cutting depth in relation to width, i.e. resistance decreases if lower cutting depth and higher width apply. Resistance is slightly changed with change of blade angle (β), increase of cutting angle above $\alpha = 45°$ is followed by high increase of resistance. Blade can be in shape of cone, semicircular, rectangular, etc. The best results provides arched circular blade with convexed frontal part under the angle of 12–15°. Tool blade enters the soil quickly, pressure spreads to the tool sides, and soil prism is formed in front of the rotor. Used and blunt blades could increase digging resistance up to 30 %. At digging velocities (0.5–2.0 m/s) resistance change is negligible, but with the double amount of velocity increases resistance significantly. Soil layer cutting depth is only value that could be counted on, when decrease of all resistances is concerned, but this decreases working efficiency.

For hard and rocky soils, blade is strengthened using additional teeth, which are taking over the initial load and are scattering the soil. Teeth are decreasing cutting resistance for 10–30 % and are protecting tool blade from wearing out too quickly. At soft soils of categories I and II, teeth influence is insignificant, because it increases friction resistance and digging resistance. Soil sticks to teeth and tool,

Fig. 2.24 Tiller working tool shape, wedge blade with "ears" (MV-4)

Fig. 2.25 Verification of a demining machine with tiller (Reproduced with permission from CROMAC-CTDT)

obstructing soil powdering and dumping. Teeth are mounted at the end of the discs for soil scattering and destruction of mines. Teeth can be made of abrasive resistant steel. Teeth or bits are mounted onto the holder of the demining roller. Bit material must be abrasive resistant to withstand the impact loading while cutting in the category of hard soil. Use of tungsten carbide has partially solved the problem and improved bit life.

Working tool should use adequate rotation speed and should be provide with adequate power for soil digging, in order to overcome digging resistance, Fig. 2.24. Usually, shape of cutting tool is adjusted according to soil category. Required working power should be adjustable and verified, due to uneven increase in digging resistance on the tiller, Fig. 2.25.

Power Required for Machine Work

$$P_r = R_{1u} v_o \ [\mathrm{W}] \tag{2.50}$$

v_o circumferential rotor velocity
Power Required for Machine Movement
Machine movement resistance

$$R_u = R_k + R_i + R_\alpha \ [\mathrm{N}] \tag{2.51}$$

$R_k = G\, f_k\, cos\alpha,$ rolling resistance (wheels/tracks)
$R_i = m\ a,$ inertia resistance ($G\ v/g\ t_s$)
$R_\alpha = G$ sin α, acclivity resistance
v machine speed
G machine mass
f_k rolling resistance coefficient
a machine acceleration (t_s machine acceleration time)
Power required for machine movement

$$P_v = R_u v \ [\mathrm{W}] \tag{2.52}$$

Total power required for demining machine

$$P_T = P_r + P_v \ [\mathrm{W}] \tag{2.53}$$

Engine power should be increased for the losses in transmission (η_p), engine cooling system, hydraulic oil cooling system, air filtration, power for auxiliary devices, etc.

$$P_m = \frac{P_T}{\eta_p} \ [\mathrm{W}] \tag{2.54}$$

2.7 Demining Machine with Rollers

Demining rollers were first used by armed forces. AP and AT mines are destroyed by direct pressure of the roller onto mine fuse. Device with discs, for AT mine activation, mounted in front on the tank is well known. By the use of discs, higher speed in opening of the passages through the minefields can be achieved. After that, machines of miller and tiller type can be used for complete soil treatment to a certain digging depth. In humanitarian demining, demining machines with rollers are used, designed and adapted for real demining conditions. For example, on very dry terrains, use of tiller is causing sand clouds, which prevents machine demining. In these conditions, devices with demining rollers can be used.

Through the analysis of efficiency of one disc device, relevant parameters for design of humanitarian demining devices can be determined. Criteria of disc durability in minefield and significant factors of reliability of buried mines destruction are identified. The most important factors are quantity of explosive, machine movement speed, soil type, and soil profile and condition.

2.7.1 Heavy Mine Rollers

Mine explosion is caused by disc pressure on mine fuse. Explosion pressure of AT mine ejects the section of discs to the height of 1.0–1.5 m [11]. When ejected, discs could be damaged. Total weight of demining device is 7000 kg. Weight of one disc is 500 kg, because AT mine are activated by pressure of 300 kg. Due to better mine detection and their activation, disc hole is larger than axis diameter for 250 mm, so they can moved in vertical plane and adjust to ragged terrain profile. *Discs rims are ribbed in order to cut through the masking soil layer and to activate mine fuse, and to ensure discs rolling—to prevent sliding that will push the mine without destroying it.* Working speed during demining is 10–15 km/h. Slope and side slope can be up to 20°. Mechanical durability of disc sections to mine explosion impacts should be around 10 explosions of AT mines (5.0–7.0 kg of TNT). Disc shortcomings are inability for quick change of direction, i.e. poor maneuverability of machine equipped with disc, difficult use on slopes, poor performance on soft soils.

Reliability of Mine Activation

When opening the passage in minefield, it is expected that under one demining disc section several mines will be activated (up to three mines). Number of destroyed AP mines in a minefield can be calculated using probability theory. Detection of AT mines and their activation using demining disc device determines reliability of mine-clearing machine operation, evaluated through two scenarios, Fig. 2.26:

$\boldsymbol{p_1}$ —probability that disc sections will hit the mine in minefield,
$\boldsymbol{p_2}$ —probability that mine will be activated by, for example, device equipped with two disc sections.

As both scenarios have to occur simultaneously, reliability of mine activation, when two disc sections ($\eta = 2$) cross over one mine lane in minefield, is $\boldsymbol{R} = \boldsymbol{p_1} \boldsymbol{p_2}$. Probability $\boldsymbol{p_1}$ that mineclearing machine will hit the mine in one minefield lane, depends on distance between two neighbouring mines, diameter of pressure sensitive mine cover, disc section width, required overlap of the mine cover and disc section. Randomly set minefields are excluded.

Demining device has to protect the vehicle it is mounted on, and probability for sections hitting the mines has to be higher than the probability of vehicle hitting the mine in one lane. It is obvious that this is achieved by width of disc sections being higher than width of wheels or tracks. Probability $\boldsymbol{p_2}$ depends on disc profile,

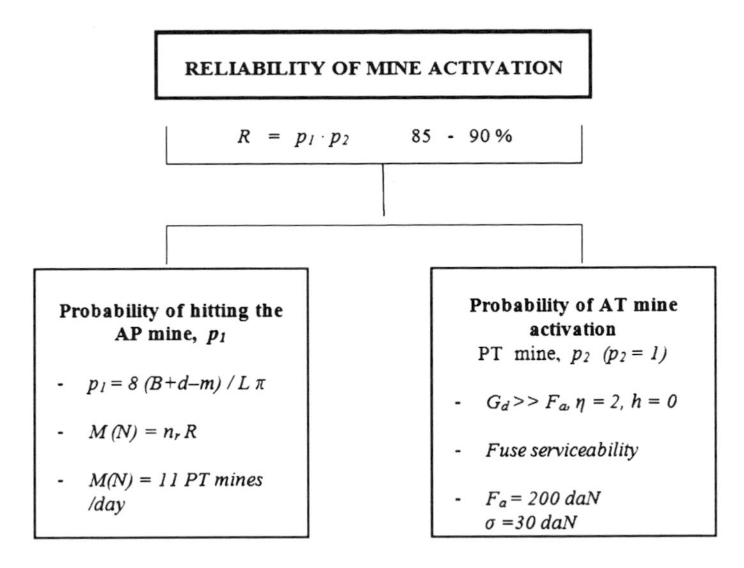

Fig. 2.26 Reliability calculation for AT mine activation

so that disc intersects the protective soil layer above masked mine and ensures direct contact for activation, $h = 0$, where h is depth of masking layer. It is assumed that the force of disc weight on the mine is always higher than force needed for mine activation, which is assured by disc weight safety factor.

Mathematical expectation for number of activated mines, when mine-clearing machine crosses minefield with "*n*" lanes, is:

$$M\,(n) = n_r R \tag{2.55}$$

Reliability of mine activation in one lane of a minefield is usually around 61 %. Expected number of activated mines M = 11 PT mines daily, with assumption that mines are laid at the common military distance of 4.5 m [11].

$$p_1 = (8/\pi)\,(B + d - m)\,/\,L \tag{2.56}$$

B width of one disc section, 0.89 m.
d diameter of AT pressure sensitive mine cover, PT mine, 0.2 m.
m required overlap of the mine cover and disc section, in order to activate the mine, 0.05 m.
L distance between two neighbour mines (from axis to axis), 4.5 m.
p_2 probability of mine activation, 1.0.
n_r number of cleared lanes per day, 18.
R reliability of mine activation for one minefield lane, $R = p_1 = 61$ %.
$M\,(n)$ mathematical expectation for number of activated mines per day, 11 mines/day.

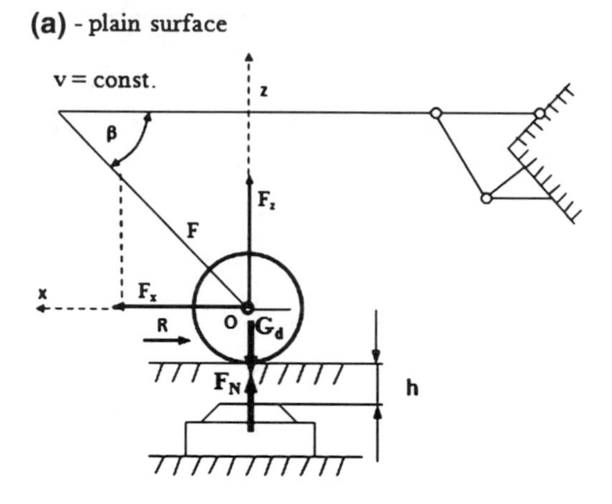

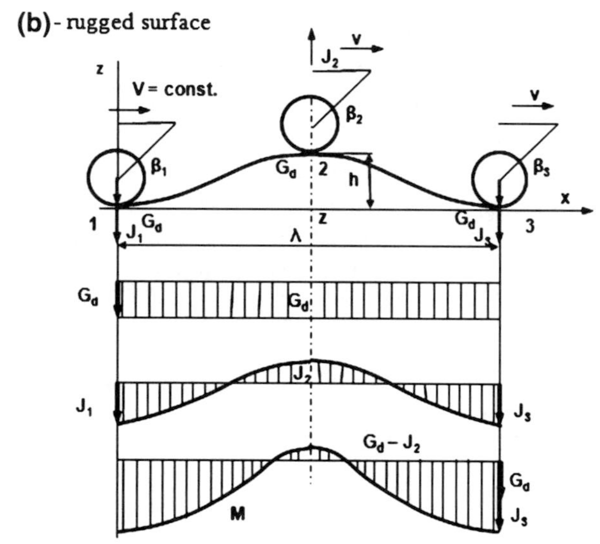

Fig. 2.27 Load at disc movement on plain (a) and rugged surface (b)

Identification of Factors for Mine Activation

According to force scheme for disc movement on ***plain surface*** and *uniform movement*, soil reaction *(F_N)* depends on soil type and disc shape, Fig. 2.27a. When disc is shaped cut through masking layer *($h = 0$)*, and when soil load bearing capability is sufficient to prevent mine from sinking under the load, soil reaction force is equal to mine activation force *(F_a)*. Discs weight has to be equal or higher than activation force, i.e. condition for mine activation is $G_d \geq F_a$.

Due to equilibrium condition, disc weight equals:

$$G_d = F_a(1 + f_r tg\beta) \tag{2.57}$$

This means that force by which disc acts on mines depends on disc rolling resistance coefficient *(f_r)* and section inclination *(β)*, not only on disc weight.

Disc movement on ***rugged surface*** can be set up along ***cos*** curve with constant movement speed, Fig. 2.27b:

$$z = h/2[1 - cos(2\pi x/\lambda)] \quad (2.58)$$

X—travelled distance, λ—period (bump length), *h*—bump height.

At disc movement on rugged surface, inertia force J_i appears, which depending on direction, decreases or increases reaction force of the surface:

$$\begin{aligned} F_N &= (G_d \pm J_i)/(1 + f_r tg\beta) \\ J_i &= M_d Z'' \end{aligned} \quad (2.59)$$

Acceleration:

$$\begin{aligned} z'' &= h/2(2\pi v/\lambda)^2 cos(2\pi t/\lambda) \\ x &= vt \end{aligned} \quad (2.60)$$

$$G_d = F_a(1 + f_r tg\beta) / 1 \pm 2h/g\,(\pi v/\lambda)2 \quad (2.61)$$

Force, by which disc affects the mine in movement on rugged foundation, depends on rolling resistance coefficient (f_r), on angle (β), on movement speed (v), on bump length (λ) and on bump height (h). Based on Eq. (2.61), disc design parameters can be calculated, as well as diameter (D) and disc width (e), Fig. 2.28. Influence of particular factors on buried AT mine activation using disc demining device is presented in Fig. 2.29 [11].

Analysis

It may be assumed that mine activation force is random value, distributed according to standard law: $F_a = F_a^{sr} \pm 3\sigma$. Mine activation force F_a is practically within limits of 1200–3000 N ($\approx$120–300 daN), mine activation diagram shows:

1. With increase in movement speed, force required for mine activation is decreasing. With speed higher than 14 km/h, mine will not be activated according to $F_N = 451\text{–}1.56\ v^2$ (daN). If tank moves at high speed, solders have to be satisfied with mine destruction reliability of 80–90 %.
2. With increase of disc rolling coefficient, $f_r = 0.12\text{–}0.45$, expressed force values F_N are adequately high for safe mine activation.
3. With increase of bump height h over 300 mm, mines are left inactivated. In order to follow the terrain profile, a gap between discs bore and axis of 160–250 mm is introduced. To decrease the influence of rough surface even more, discs rims are ribbed. Soil is cut by ribs and influence of rough surface is decreased, $F_N = 451\text{–}11.2\ h$ (daN).
4. With increase of bump length $\boldsymbol{\lambda}$ over 2 m, activation of all mines may be expected, but not below distance of $\boldsymbol{F_N} = 451\text{–}2016/\boldsymbol{\lambda}^2$ (daN). For value $\boldsymbol{\lambda} = \infty$, movement on smooth surface is established.

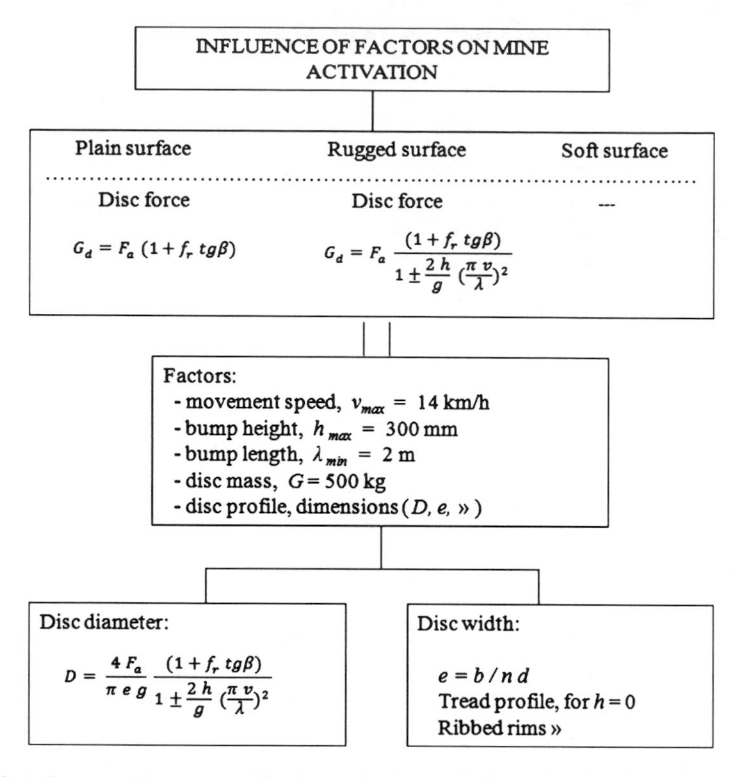

Fig. 2.28 Influence of factors on AT mine activation and disc design parameters

Therefore, *significant factors for mine activation using discs are* movement speed, surface profile, disc shape, soil conditions and inclination of disc sections.

Basic user requirements for demining machine development, i.e. demining device with discs are as follows:

Durability of device against AT mine explosions, (6 kg TNT—at least 11 mine),

Reliability of mine activation and mine destruction, $p = 80$–$90\ \%$,

Movement speed for mine removal of 10 km/h,

Crew protection against AT mine impulse noise, peak value of 150 dB,

Quick replacement of damaged discs.

2.7.2 Medium Mine Rollers

Demining machines can neutralize laid AP mines using weight of discs, which are placed in one or more roller sections. Behind machine, rakes for soil preparation can be attached in order to collect metal fragments from soil surface using magnets, before inspection with metal detectors.

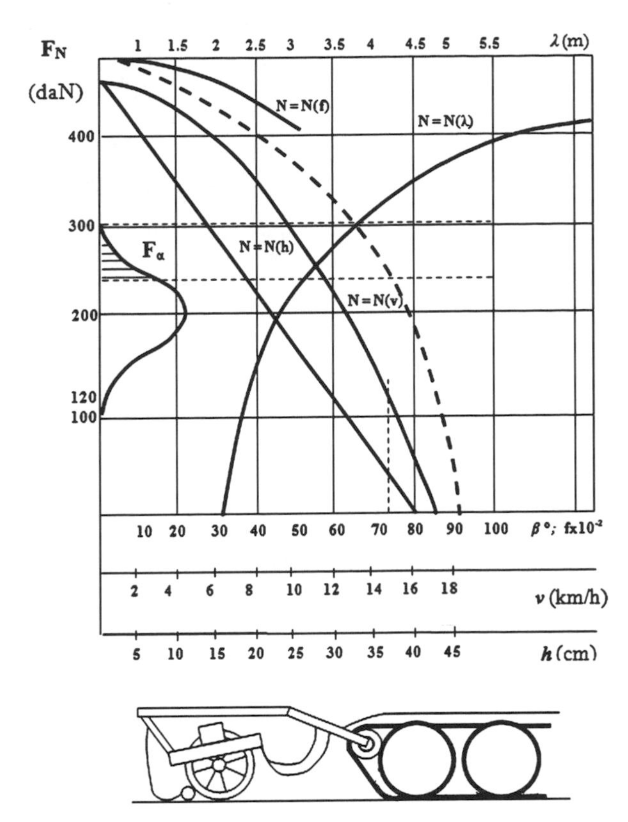

Fig. 2.29 Mine activation diagram of buried AT mine with pressure-activated fuse, using disc demining device

Machines with medium and mini rollers may serve as additional machines in machine demining, for the following purposes:

- to prepare surface for demining, to detect mine polluted areas and to reduce mine polluted areas,
- for demining the areas that are polluted only with AP mines, as part of overall demining mechanization based on mine neutralization (crushing, activation),
- for the control of mine pollution of roads for road maintenance in mine suspicious areas.

Advantages of medium and mini rollers in demining

- for determining the level of mine pollution in mine suspicious areas, larger number of rollers can be used at the same time in demining operations,
- possibility to replace rollers with other demining tools,
- collecting the metals after the use of other demining machines,
- low serial roller production costs.

Disadvantages

Due to soil intersection, especially spongy and muddy soil and massive vegetation, demining of mine polluted area by pushing the rollers in front of the

Fig. 2.30 Light demining machine with roller, 2t, 10 segments. (Reproduced with permission from Ref. [8])

machine, is aggravated or almost impossible. That's why rollers are in some cases towed behind the mine resistant machine.

Accidental AT mine activation can damage the rollers, although well designed discs provide possibility of blast ventilation.

It can be concluded as follows:

demining machines with medium and mini rollers can be used in humanitarian demining as a part of entire mechanization for machine demining as well as for road maintenance in mine suspicious areas (Fig. 2.30).

References

1. Humanitarian demining - Requirements for machines and conformity assessment for machines (2009) Standard HRN 1142, Croatian Standards Institute, HZN 1/2010, Zagreb.
2. Nell S (1998) Experimental Evaluation of Mean Maximum Pressure (MMP) using Wheeled Vehicles in Clay, Wheels & Tracks Symposium, Cranfield University, Royal Military College of Science, Cranfield.
3. Sarrilahti M (2002) Soil Interaction Model, Project deliverable D2, Appendix No.2. Development of a Protocol for ECOWOD, Univesity of Helsinky, Department of Forest Resource Management, Helsinky.
4. Mikulic D, Marusić Z, Stojkovic V (2006) Evaluation of terrain vehicle mobility, Journal for Theory and Application in Mechanical Engineering 48, Zagreb.
5. Mikulic D (1998) Construction Machines, Design, calculation and use (in Croatian), Zagreb.
6. Shankhla V S (2000) Unravelling flail-buried mine interaction in mine neutralization, DRES TM 2000-054, Defence Research Establishment, Suffield.
7. A Study of Mechanical Application in Demining (2004) GICHD, Geneva.
8. Vinkovic N, Stojkovic V, Mikulic D (2006) Design of Flail for Soil Treatment, 5th DAAAM International Conference on Advanced Technologies for Developing Countries, University of Rijeka, Rijeka.
9. Suceska M (1999) Calculation of detonation energy from EXPLO5 computer code results, Propellants, Explosives, Pyrotechnics 24/1999.
10. MV-4 Mine Clearance System (2012) Catalogue, DOK-ING Ltd, Zagreb.
11. Mikulic D (1999) Demining techniques, modern methods and equipment, Demining Machines (in Croatian), Sisak, Zagreb.

Chapter 3
Design of Demining Machines

Abbreviations

A_i [%]	Technical availability
MTBF [h]	Mean Time Between Failure
MTTR [h]	Mean Time To Repair
Λ [-]	Failure rate
$M(t)$ [h]	Mathematical probability
R [m]	Flail radius
F_i [N]	Force impulse
R_{ki} [N]	Hammer cutting resistance
$R_{\sigma\ i}$ [N]	Resistance to crushing
P_T [W]	Total power for machine
P_r [W]	Power for machine operation
P_v [W]	Power for machine movement
Ps [W]	Flail specific power
M_{hm} [Nm]	Hydromotor moment
M_g [Nm]	Moment for starting the track movement
M_u [Nm]	Rotor resistance moment
M_p [Nm]	Flail start up moment
R_I [N]	Soil cutting resistance
$F_{\sigma i}$ [N]	Total crushing force
S [mm]	Tool cutting feed (hammer shear)
S_{ti} [m]	Thickness of crushed
z_n [-]	Number of grasping hammers
R_k [N]	Rolling resistance (wheels/tracks)
R_i [N]	Inertia resistance
R_α [N]	Slope resistance
$\Sigma\ R$ [N]	Movement resistance

D. Mikulic, *Design of Demining Machines*,
DOI: 10.1007/978-1-4471-4504-2_3,

f_k [-] Rolling resistance coefficient
M_{gi} [Nm] Track moment
r_g [m] Track sprocket wheel radius
P_{DE} [W] Diesel engine power
U [m^3/h] Machine efficiency

3.1 Project Requirements

Requirements for demining machine development include basic requirements and requirements for design of particular systems. Requirements for machine development are based on needs for demining in hardest specific conditions. Designers are often required to develop machines that should fulfil specific requirements. Requirements arise for light transport, demining in hard terrains, at high temperatures in deserts, and under heavy dust. For example, use of machines in dry terrains causes heavy cloud of dust. This is the result of crushing the surface layer when treating the soil. Big and dense sand cloud influences machine operation: machine can operate only in up-wind direction. Additionally, there is a problem of air filtration at the drive engine air-inlet, causing machine to stop the operation, thus decreasing machine efficiency.

Basic and specific requirements become every day more and more rigid. Most of demining machine design requirements is achievable, and process of humanitarian demining is more and more mechanized. Mine neutralization is done quickly and safely for deminers. Basic requirements for speeding up the demining process demand for better design of demining machines and additional demining equipment. Application of new technologies is frequent, and machine designers and manufacturers are constantly improving machines according to new requirements.

Basic requirements for demining machine development are:

- Defining of machine category in accordance with efficiency requirements in specific conditions of use: light, medium and heavy demining machines.
- Defining of demining machine structure, tracked vehicle or wheeled vehicle, working tool (flail, tiller, roller, cutter).
- Demining of mine obstacles (AP, AT and combined minefields) in real conditions (roots, stones and other obstacles). Removal of explosive, ammunition and mine remains, which could be potential threat.
- Machine should neutralize mines by crushing or by detonation. Tool should be resistant to wear-out, mine explosion, roots and stones. Tool teeth should be replaceable.
- Machine should operate at technological speed adjusted for soil treatment and certain digging depth of 10–20 cm.
- Mine clearance quality should be according to standard IMAS [1].
- Demining cost per square meter should be less than cost of manual clearing of big areas,

- Machine design should have: device for remote control, and GPS guiding system,
- Machine should provide high working reliability in minefield, excluding possibility of so called "mouse trap", i.e. machine should pull itself out using additional engine,
- Machine life cycle is 5,000–10,000 working hours, preventive maintenance included, and no replacement of vital assemblies. Machine may be destroyed due to mine explosion and become irreparable.
- Machine crew should be protected against AP and AT mine fragments, impulse noise and vibrations. Machine should be armoured from all sides, especially against sudden explosions from the front and underneath.

  - Basic machine testing includes:
  - Machine crew safety test,
  - Determination of performance/mine clearance, efficiency,
  - Machine durability test and maintainability,
  - Logistics acceptability.

Project requirements can be put into groups:

Demining machine structure
Operating conditions
Performance requirements
Design requirements
Reliability requirements
Documentation requirements
Testing requirements
Evaluation model

Requirements for the machine design depend on the demining machine category (light, medium, heavy). Machines can be remotely controlled, tracked or wheeled and designed for clearing AP mines and, if possible, AT mines. Machines need to be suitable for clearing between buildings, along paths, in plantations and other areas where the ground cannot sustain heavy loads. Machines need to be easily transportable with terrain trucks and by helicopter.

Operating digging tool of the mine clearing unit needs to be a flail mounted in front of the machine, which destroys the smallest antipersonnel mines and the most dangerous antipersonnel bouncing mines. Mine neutralization is being performed by the force of impact tools of appropriate shape—hammers at the end of the flail chains. Flail rotation and hammers strokes are digging the ground down to the depth of twenty centimetres, together with mine destruction or activation. Machine has to have very high performance in soil processing in accordance with humanitarian mine clearance technology.

Flail operation is based on flail impact force (striking on chain), i.e. impact moment at which flail is penetrating or crushing the soil. Surface soil layer is scattered towards machine shields, which levels it. The distance of the flail rotor from the explosion centre is equal to flail rotation radius. Due to chain connection

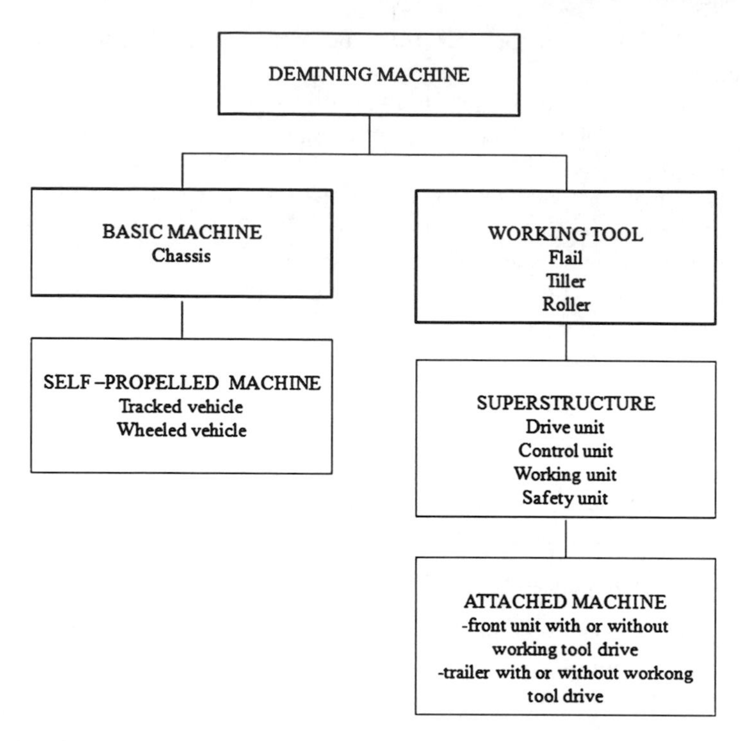

Fig. 3.1 Demining machine structure

between rotor and striking hammers, flail rotor is not directly stressed by the soil digging resistance moment, and less power is estimated as required for flail operation in comparison to tiller operation.

3.1.1 Demining Machine Structure

Demining machine consists of basic machine and working tool for demining, Fig. 3.1. Basic machine can be on tracks or wheels. Working tool for demining can be flail, tiller and roller. For removing vegetation, a cutter is used as a working tool. Machine development can be of original construction or an adjusted construction machine. Modern demining machines use hydrostatic drive for machine movement and rotation of working tool.

3.1.2 Operating Conditions

In mine-suspected areas, AP and AT mines and UXO can be laid. Minefields can be old or recently laid. Mines are of different types, from small pressure fused AP mines, with or without scatter function, to tripwire fused bouncing-fragmentation mines.

Demining machine needs to operate in the top layer of various grounds, to the required depth. Demining machine must operate in all climatic and weather conditions, except for mine clearing in frozen ground. Temperature range when the system shall have full function is between −5 °C and +45 °C. Demining machine shall operate in all kind of terrain with brushwood up to 3–5 cm in diameter and inclinations up to 15–25 %.

3.1.3 Performance Requirements

- Light demining machine (LDM) shall clear all known pressure and tripwire fused anti personnel mines (and antitank mines).
- LDM must clear mines to a depth of 20 cm, measured from ground level to the bottom of the mine.
- LDM must clear at least an area of 500 m^2/h.
- LDM must have minimum specific power of 75 HP per meter of toll (flail) width.
- LDM maximum mass 5,000 kg (5 t).
- LDM have to be equipped with track unit with drive gear.
- LDM must clear mines on slope (upward and downward) of 25 %, and side slope of 15 %.
- LDM must be fully operational in vegetation like brushwood up to 3 (5) cm in diameter, with relative distance of 10 cm between trees.
- Mine clearing unit must clear an area 20 cm wider than the machine, on each side, to ensure safe maneuvering.

Mobility and transportation

- LDM must be transportable by truck or a trailer.
- LDM must be air transportable by CH-47 helicopter (7 t payload).
- LDM must be operational within 10 min after being downloaded from the truck or trailer.
- LDM must be able to maneuver at slopes of 30 % (downhill and uphill) and side slope of 25 %.
- LDM must be able to cross a ditch 0.5 m wide and 0.5 m deep.
- LDM must be able to move over vertical obstacle of 0.3 m high.
- LDM must be capable for fording of 0.3 m deep water without any special preparations.
- LDM must be able to stay in full clearing mode for five hours without refueling and maintenance.

Remote control

- LDM must be able to operate remotely controlled from a distance of 200 m (500 m) within direct line of sight.
- Operator must be able to remotely operate the LDM from the inside of an armoured personnel carrier or similar armoured vehicle.
- Remote Control system should use radio communication between the LDM and the remote control unit.
- Operator should be able to read LDM operational parameters on the Remote Control unit display (e.g. oil temperature, oil pressure, engine temperature, fuel, rpm of the clearing unit etc.).
- Remote Control unit must have a warning system/signalization in order to prevent fatal damage to LDM vital systems and parts.
- If communication between the operator and the vehicle is interrupted, the vehicle and the clearing device should stop operation.
- Remote Control Unit must have a built-in-self test system to ensure proper function.
- Remote Control Unit must provide the operator options to use fixed, pre-programmed clearing speeds (e.g. 0.2 km/h, 0.4 km/h, 0.6 km/h, etc.).

Soil digging density and depth profile

Clearing the ground needs to be achieved by the cutting feed of the 16–30 mm hammer. A realistic option of hammer feed is 30 mm which is in fact half of the diameter of the smallest AP mine, while the same mine's fuse length is 16 mm.

To achieve a soil digging density and depth profile, a diagram of *machine speed—flail revolutions per minute for determined depth* (hammer hit distance) needs to be defined.

Mine-clearing

- The mine clearing probability in the first pass should be 100 %, for all known types of mines.
- While clearing mines, the LDM must not scatter unaffected mines to the side, beneath or behind itself.
- No special preparation by the crew is required for the LDM towing.

3.1.4 Design Requirements

- LDM must be designed to fulfil all operational requirements in all climatic and terrain environments for the entire lifetime of the vehicle.
- LDM should consist of a main machine-chassis and an exchangeable mine clearing device.
- Engine must be of commercial type in serial production.
- Engine must be able to operate on both diesel and NATO F-34 fuel (kerosene).
- Engine must have full monitoring.

- Power pack active protection—automatic fire extinguishing system
- The LDM must operate using standard lubricants and fluids.
- Fuel tank interior must be protected against corrosion and influence of fuel.
- Cooling systems must be fully functional and allow mine clearing operations in dusty environment for at least 5 h at temperature range from −5 °C to + 45 °C.
- Cleaning of cooling-systems from dust build-up should not take more than 15 min.
- Air intake-systems must be fully functional and allow mine clearing operations in dusty environment for at least 5 h at temperature range from −5 °C to +45 °C.
- Air—intakes must be placed high enough to minimize the risk of the intake blockage by ice, leafs etc.
- Air—intake systems must have a device to prevent larger particles from blocking the filters.
- Hydraulic system must use hydraulic oils that will ensure long-term operation with no need for hydraulic fluid change.
- Hosing must be protected against mechanical wear-out.
- Hydraulic systems must be protected from fragment impact originating from mine-clearing action.
- Mine clearing unit must be designed for easy maintenance i.e. bolts, nuts and other connections should be minimally affected by the mine-clearing action.
- Elements or device for the prevention of damage caused by wire winding must to be incorporated in the rotor of the clearing unit.
- Machine must have proper towing connectors that are easily accessible in all different towing situations.
- Machine must have drains for evacuation of fluids (e.g. water, oil, hydraulic fluids, etc.) from the engine and other compartments.

Endurance and reparability

- LDM vital parts must withstand repeated detonations from AP-fragmentation mines (as PROM-1) from the distance of 5 meters.
- LDM must repairable after an AT mine detonation under the mine clearing unit.

3.1.5 Reliability Requirements

Technical availability: describes the probability that system or unit will be operational for a specific time period, providing that the regular maintenance has been performed (inherent availability).

Technical availability, A_i

$$A_i = MTBF / (MTBF + MTTR) \quad (3.1)$$

MTBF Mean Time Between Failure
MTTR Mean Time To Repair

Functional Reliability: describes the probability at which a system or unit will operate satisfactory without repair if it is used properly in the environment for which it was designed. Functional reliability is measured in terms of mean time between failures *(MTBF)* or failure rate *($\lambda = 1/MTBF$)*. In this case, *MTBF* is the time in operation.

Maintainability: describes the probability of repairing or maintaining the system or unit in the shortest possible time in order to keep the unit operational. Maintainability is measured in terms of mean time to repair *(MTTR)*.

Durability: describes how the overall condition and operating capabilities of the system or unit are changing during its life cycle, if it was used and maintained properly.

Requirements in demining

Availability: (A_i) for the system must be at least 0.90 (90 %) during operations.

Functional Reliability:

Mean time between failures *(MTBF)* for failures that are causing the machine to stop the operation, should be at least 100 h.

LDM must resist detonations as follows:

- explosive weight of 0.1 kg—no need for repairs or corrective maintenance.
- explosive weight of up to 0.5 kg—mine clearing unit should continue normal operation. Repairs or corrective maintenance in this case (repair or replacement) must be performed within 30 min.
- explosive weight of up to 1.0 kg—mine clearing unit may be damaged. Repairs or corrective maintenance in this case (repair or replacement of repairable units) should be performed within 4 h, otherwise replacement of complete mine clearing unit at the operation site is recommended.
- for the explosive weight of 6.0 kg—mine clearing unit may be complete replaced.

Maintainability:

- Standard tools must be sufficient for maintenance and repairs.
- Operator alone should be able to replace parts at the first level maintenance.
- Operator and support personnel should be able to replace the mine clearing unit within 4 h.
- Maintenance personnel should be able to replace the mine clearing unit on site.
- Maintenance personnel should be able to perform major repairs tasks in workshop.

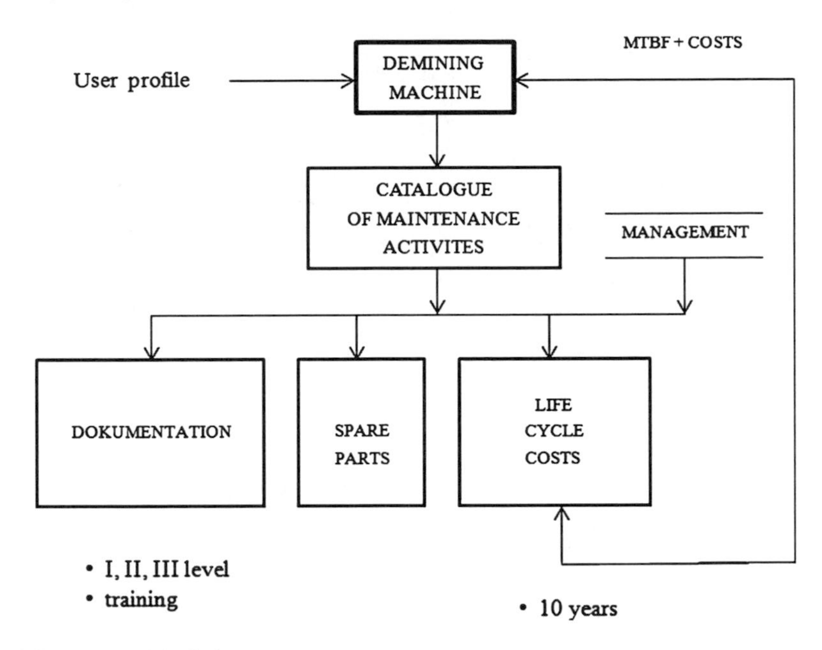

Fig. 3.2 Integrated logistic support

Operator alone should be able to perform maintenance tasks as follows:

- before operation within 20 min
- after operation within 30 min
- special inspection within 4 h

Machine durability

Machine life cycle is limited due to repair costs and low reliability level. Low reliability and high maintenance costs are basic criteria preventing uneconomic machine overhauls. Machine durability till medium overhaul can be calculated using mathematical expectation:

$$M(t) = \Sigma t_i \, g_i \, [\mathrm{h}] \tag{3.2}$$

t_i mean time between failure *(MTBF)*

g_i price share of repairs for particular assembly in relation to total machine price

C_i/Cu including direct costs (spare parts, time required for repairs)

Calculated machine durability, calculated in machine working hours, till medium overhaul is realistic limitation for which reliable machine operation can be anticipated. At medium overhaul, smallest key assemblies of hydraulic device should be replaced. Integrated logistic support of a demining machine is given on Fig. 3.2.

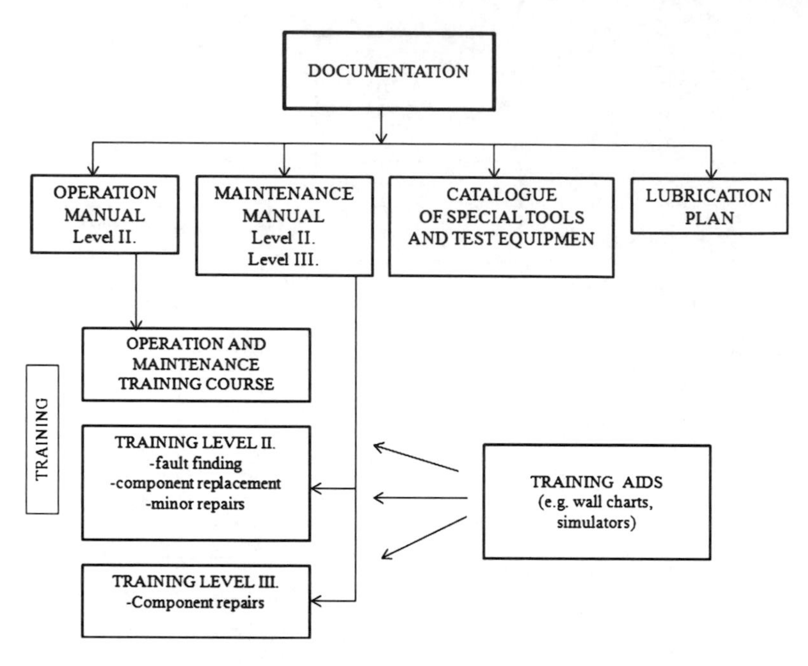

Fig. 3.3 Documentation requirements

3.1.6 Documentation Requirements

Documentation includes all manuals, drawings and other documents that provide necessary information for education, operation and maintenance work for demining system, Fig. 3.3.

Contractor of demining system must provide the following documents:

- Operator's manual
- Maintenance schedule for regular maintenance
- Maintenance schedule for special maintenance
- Lubrication schedule
- Spare part catalogue
- Technical manual

3.1.7 Testing Requirements

- The light demining machines must be tested according to the CWA [2, 3].
- LDM must be tested for vibration according to MIL-STD810 C, and must be tested for shock according to MIL-STD810 E.
- Climatic: LDM must be fully operational within temperature range from −5 °C to +45 °C.

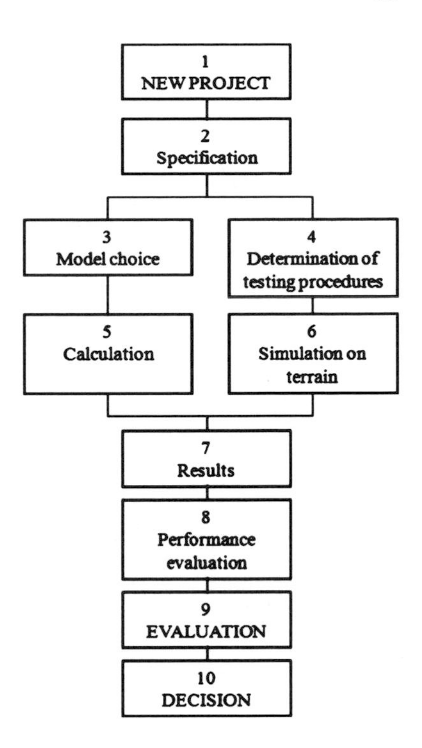

Fig. 3.4 Demining machine evaluation model

- LDM and subsystems must withstand ESD, radio and radar transmission generated fields, power and telephone line transients whether caused inductive coupling or voltages, or equipment generated voltages.
- Working environment. Eventual dusty environment must be considered in all design of the light demining machine.

3.1.8 Evaluation Model

When designing the machine, the most important factor is its use in humanitarian demining, Fig. 3.4. Specification description is determined by customer requirements, technical specifications, regulations, etc. (2). Calculation results (5) and prototype simulation (6) are compared to basic specifications (2). Based on achieved results (8), real machine performance is evaluated. Then, it is possible to evaluate machine design or certain options (9), according to CWA [3]. Results of this evaluation can provide for final decisions on design (10).

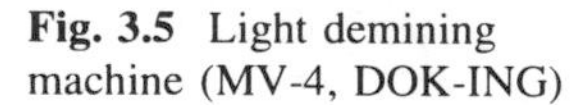

Fig. 3.5 Light demining machine (MV-4, DOK-ING)

3.2 Design of Light Demining Machine

The demining machines have been developed for mine clearance of mine suspected areas, mined house surroundings, and terrains difficult to access by bigger machines. Strict requirements have been set during development of the light 5 ton machine (Fig. 3.5), such as: safe clearance of antipersonnel mines, intensive use in most severe working conditions, ability to operate on all soil categories, work in extreme summer conditions of temperatures; high machine efficiency; safety of operators during operation; modular design and lowest possible maintenance requirements; ballistic protection and remote control. Antipersonnel mines are neutralized by the force of impact tools of suitable shape—hammers at the end of the flail chains. By flail rotation and hammer hitting, the loose soil layer of up to 20 cm deep is dug and the mines are crushed or activated.

3.2.1 Soil Digging

Soil digging is based on the force impulse of the flail hammer of 0.6–1.0 kg of weight. Depending on the soil moisture, hammering cuts the soil and throws it toward the guard. If the soil is dry and hard, then it is crushed and dispersed towards the guard. Universal "mushroom" hammer shape is used. Cutting can be done on the coherent soil, well bound and soft. Crushing can be done on the soil which is non-coherent, dry and hard. In the former case, the calculation is applied according to the soil cutting theory equations, and for the latter case the criteria of the boundary soil strength σ is applied. Hammer operation principle of overcoming the soil cutting resistance is based on the force impulse, Fig. 3.6. In order to realize the procedure of cutting the condition that the force impulse of the hammer is greater than the digging resistance has to be fulfilled.

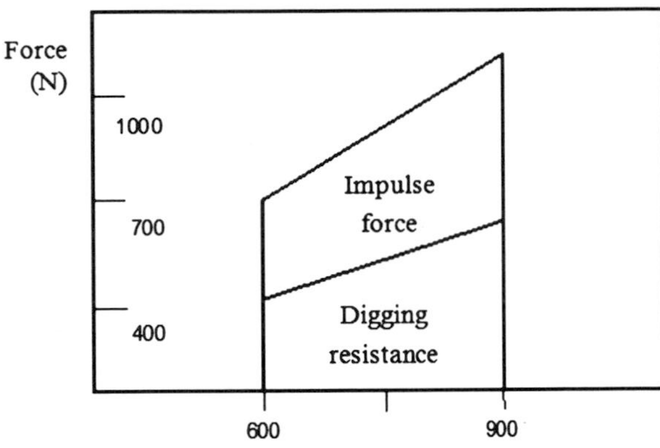

Fig. 3.6 Impulse force of the light flail hammer and digging resistance

Impulse force of the flail hammer:

$$F_i t = m_h v_o$$
$$F_i = m_h v_o / t \tag{3.3}$$

Cutting resistance:

$$R_{ki} = z_n k_1 b S_t \ [\text{N}] \tag{3.4}$$

Crushing resistance:

$$R_{\alpha i} = \alpha A \ [\text{N}] \tag{3.5}$$

m_h hammer mass, t layer cutting time,
v_o hammer circumferential velocity, z_n—number of hammers in contact,
k_I specific resistance of soil cutting, according to soil type
b hammer cutting width,
S_t current thickness of the soil cutting layer ($S \sin\varphi$)
S hammer cutting feed providing soil digging density for safe mine neutralization,
A area of the hammer impact blade, σ—soil strength (non-coherent soil)
h soil digging depth

3.2.2 Drive System

Armoured body accommodates a light Diesel engine which, via a gearbox, drives the variable-flow hydropumps for machine operation and movement, Fig. 3.7. The flail operating device contains one supply hydropump in the hydrostatic closed circuit, two hydromotors and two reduction gear chains connected to the flail rotor. Drive is also supplied to the pump which extends/retracts the telescope flail arms.

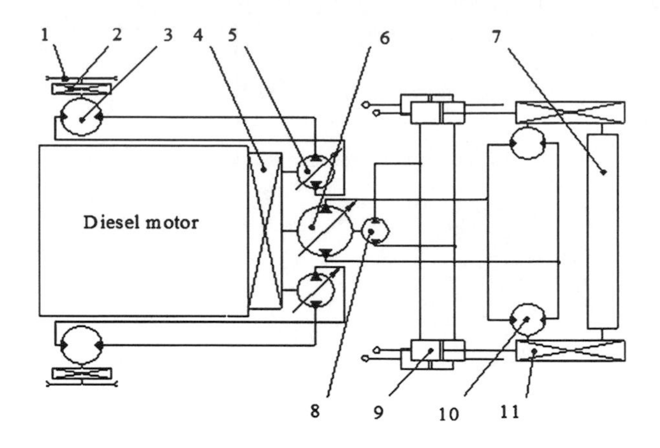

Fig. 3.7 Hydraulic scheme of a demining machine (MV-4) *1* drive chain wheel, *2* planetary reduction gear, *3* track hydromotor, *4* gearbox multiplicator, *5* drive hydropump, *6* flail hydropump, *7* flail rotor, *8* hydropump for telescope flail arm, *9* hydrocylinders for telescope flail arm, *10* flail hydromotor, *11* chain

Track drive has two independent closed hydrostatic circuit, one circuit per track. Planetary reduction gears of high transmission ratio (i_{pr}) are mounted between the hydromotor and the tracks.

3.2.2.1 Engine Power and Hydraulic System

The total flail rotation resistance moment includes static and dynamic moment of flail elements rotation, until the moment when hammer impacts the soil or until the loss of kinetic energy, change of angular speed and reduction of the flail rpm. Since this phenomenon of energy exchange is not completely understood in practice, the starting assumption is the cyclical operation of each flail: hammer acceleration, and almost stopping at cutting or crushing the soil layer.

Required machine power consists of the power for operation and the power for machine movement: $P_T = P_r + P_v$. The amount of power required for transmission, engine cooling, hydraulic oil cooling, air cleaning, for secondary devices drive, etc. should be added to this power. Power required for machine operation and movement has been calculated, and the in total amounts to 110–150 kW (150–200 HP). Out of this, more than 90 % of power is used for soil digging and the rest for the machine movement. An adequate Diesel engine has been selected which, via gearbox, drives hydropumps for machine operation and movement. The variable-flow hydropumps have been selected, as well as the piston-axial hydromotors in their optimal oil pressure area (Fig. 3.8).

Selection of hydromotor and hydropump for flail operation

Hydromotor moment for flail operation:

$$M_{hmm} = M_k / 2\, i_{lr} h_{lr}\, [\mathrm{Nm}] \tag{3.6}$$

This is followed by the selection of the hydromotor, nominal capacity, flow, and other characteristics (catalogue). As one hydropump supports two hydromotors, then: $g_{hp} = 2\ g_{hm}$. i_{lr}—transmission relation of the reducing gear—chain, η_{lr}—utilization level of the chain wheel, g_{hp}—spec. flow (l/min)

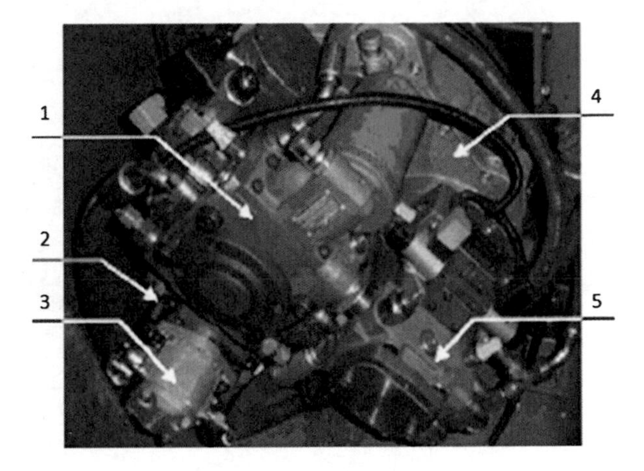

Fig. 3.8 Hydraulic pump system; *1* flail hydropump, *2* drive hydropump (*left track*), *3* telescope flail arm hydropump, *4* gearbox multiplicator, *5* drive hydropump (*right track*) (*Reproduced with permission* from Ref. [10])

Selection of hydromotor and hydropump for machine movement
Moment for starting the track movement:

$$M_g = P_v / 2\,\omega \text{ [Nm]} \tag{3.7}$$

P_v power for machine movement

Nominal moment of track hydromotor:

$$M_{hm\,g} = M_g / i_{pr} h_{pr} \text{ [Nm]} \tag{3.8}$$

After that follows the selection of hydromotor, nominal capacity and other characteristics. Hydromotor and hydropump of appropriate power has been selected. Hydraulic fluids formulated with premium anti-wear hydraulic additives can provide the benefits of durability and performance retention to meet the stringent demands.

The hydraulic system sourced from a single feed tank and cooled by a fan cooled coil, located below the forward hatch of the machine, Fig. 3.9. The hydraulic capacity of the reservoir is 200 *l*. The hydraulic tank contains an independent filter for each hydraulic circuit. The four independent hydraulic circuits control the working tool, right and left drive trains and the telescopic arm. Each circuit can be isolated from the feed tank independently in case of failure or leakage without draining the entire reservoir, to recover or repair the machine. The drive train circuit can be bypassed by enabling bypass valves located interior to the engine compartment to remove flow to the drive train to allow towing of the machine and allowing the tracks to move freely. Cooling for the hydraulic system is accomplished by a 9.5 *l* coil that has two thermally activated cooling fans. The coil can be positioned vertically for maintenance and lays flat during machine operations.

The hydraulic circuits are pressurized using engine torque to drive the multiplicator assembly which drives four hydraulic pumps. These pumps are controlled

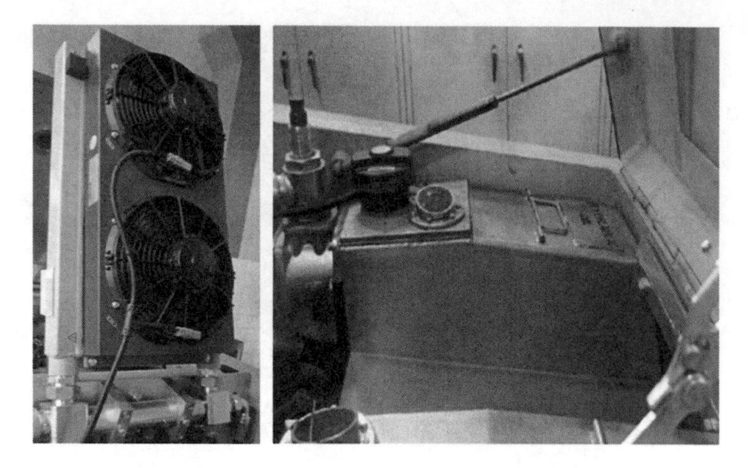

Fig. 3.9 Hydraulic oil cooler and reservoir, maintenance position (MV-4)

through the machine remote system to control flow direction and pressurize various motors and valves throughout the system. The machine boom control is accomplished by valve bank located in the electronics compartment of the machine. The valves can either be controlled through the remote system during normal operation or through the series of levers, one for each valve. These valves control the lift and lowering of the vehicle boom, and the extension and retraction of the machine arms. Located within the same compartment is the weight reduction valve. This valve is controlled through the remote system and reduces the weight of the actual tool pressure being applied to the ground. The effective maximum pressure reduction is 50 % of actual tool weight. This valve, when used with the machine float valve, ensures proper ground penetration regardless of terrain change.

The drive train utilizes hydraulic pressure to propel the machine forward, reverse, left, and right. The operator can select one of ten gears to travel at various speeds during operations. The drive train is made up of two independent hydraulic drive circuits, an automatic transmission and track tensioning wheel. Hydraulic pressure applied to the drive motor applies energy to the transmission, where it is converted to torque to propel the machine. The tensioning wheel ensures free movement of the track and prevents binding during operation and is adjustable to compensate for track wear and stretching. The transmission propels the machine by rotating the main drive sprockets exterior to the machine, Fig. 3.10. The sprocket is engaged with the track to transfer the transmission torque to track movement. The track rides on numerous road wheels located the length of the track to reduce resistance and maintain track alignment. The remote system controls pressure and direction of the hydraulic fluid to the motors which determines direction of travel and speed. In order to turn the machine, one track is locked or counter rotated in reference to the other to force the machine to pivot in its position or along a path.

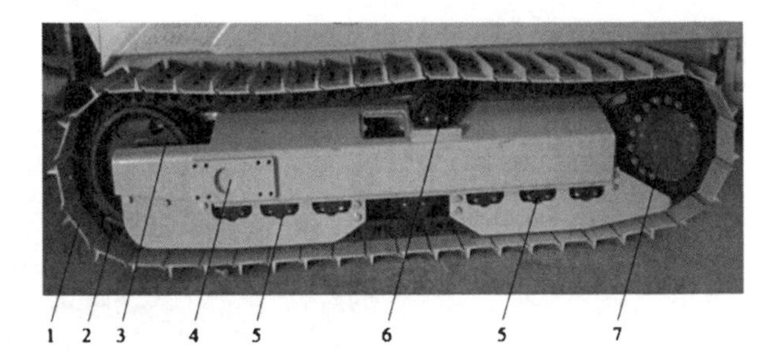

Fig. 3.10 Track model (MV-4) *1* grouser plate, *2* track chain, *3* idler/front guide, *4* tool mount, *5* road wheels, *6* support roller, *7* drive sprocket and hydromotor

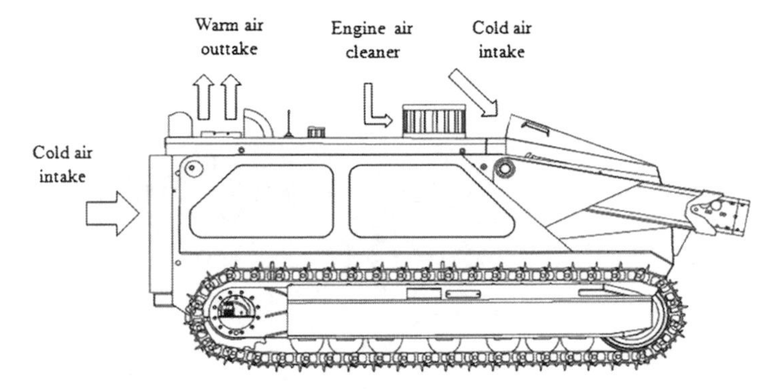

Fig. 3.11 Air stream, in function of motor and transmission cooling and air intake (MV-4)

3.2.3 Cooling System and Air Cleaning

3.2.3.1 System Cooling of Engine and Hydrostatic Transmission

Hydraulic oil for hydrostatic transmission is cooled through oil cooler and ventilator. For calculation, the input–output oil temperature is important, as well as air temperature, oil flow in the cooler, air flow through the cooler and the amount of heat that should be dissipated by the cooler. The oil that flows through hydropumps drain ducts and hydromotors is also cooled. Due to high requirements for machine operation, the energy balance is calculated based on the thermodynamics equations, and the average oil temperature at system component outlets is determined and based on optimal temperature at the cooler output. Required specific cooling capacity, oil flow through the cooler has been determined, and the adequate cooler is selected.

Air intake for cooling of oil hydrostatic transmission is located at the front of the machine. Here, hydraulic oil is cooled by means of oil cooler with two electric ventilators. Air intake for engine cooling is located at the rear part of the machine, Fig. 3.11. Behind the wire cleaner, a split radial ventilator, driven by the Diesel motor, is situated. One side of the ventilator cools motor fluid through a water cooler, followed by a vertical throw-out of hot air with dust. The other side of the

Fig. 3.12 Demining in conditions of high dust concentration in the air (Reproduced with permission from DOK-ING)

Fig. 3.13 Location of the two stage air cleaner (Reproduced with permission from Ref. [10])

radial ventilator provides longitudinal air flow, by which it intakes air from the front part of the machine and throws out hot air and dust from the machine interior, likewise in the rear part of the machine. By uptake air flow on the front side of the machine and throw-out air flow on the rear side of the machine, air cleaner area is protected from high dust concentration. Overall air stream is, therefore, in the function of motor and transmission cooling and clean air intake towards the cyclone air precleaner located on the machine roof (Fig. 3.12).

Dust concentration, rough approximation:

Constructions machinery: 35 mg/m^3, 1–1.5 m above the ground
Demining machines: 35–100 mg/m^3, 1–1.5 m above the ground

Requirements for the air cleaner:

- Air cleaner has to selected as of modular type, according to extreme conditions of dust concentration and temperature range from −40 °C to +80 °C
- Air cleaner needs to be place as close to the motor as possible

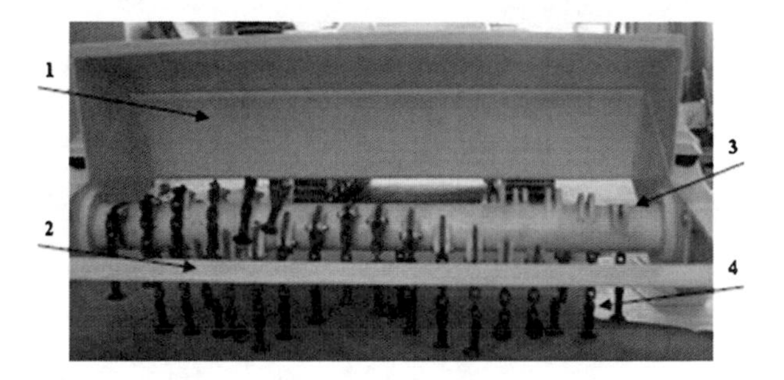

Fig. 3.14 Flail system (Reproduced with permission from DOK-ING) *1 protective shield, 2 protective bar, 3 tool shaft-rotor, 4 flail/chain and hammer*

- Air cleaner has to have enough space for an easy replacement of cleaner filter
- Service indicators have to be visible

The engine is a 4—stroke, 6—cylinder, electronic controlled governed turbo charged Perkins engine (5980 ccm), with a strength of 129 kW at 2200 min^{-1}.

Calculation flow rate needed for the motor to operate with maximum strength.

$V_t = 6.2\ \text{m}^3/\text{min}$

3.2.3.2 Cyclone and two Stage air Cleaner

For all heavy duty applications in dust, a combined model of dry air cleaners is selected: cyclone—air pre-cleaner and two stage air cleaner.

Cyclone is an air pre-cleaner with continuing expulsion that separates and removes the most part of airborne particles: dust, rain, insects, snow, sand, sparks, and cinders, it also replaces the rain cap. Extraction efficiency approximation 87.5 % of SAE coarse dirt and up to 99 % of larger airborne particles. Larger particles of dust, because of centrifugal force, are separated towards periphery and are thrown at the housing rim. Cyclone is a *Vortex* pre-cleaner that increases the service interval of the single stage air cleaners.

Two stage air cleaner*—modular system, MANN Europiclon,* Fig. 3.13.

Features: High dust capacities and low flow resistance

Medium and heavy duty conditions

Cleaner working in two stages to clean air, e.g. pre-separation with subsequent fine filtration.

Main and secondary elements: Filter elements with radial seal

Design: Metal free two stage air cleaner of fully recyclable plastic.

3.2.3.3 Extraction Efficiency and Service Life

The most important task of a dry air cleaner is to ensure sufficient protection of the engine against wear under all imaginable dust conditions. It is very important to indicate the filter extraction performance in relation to the particle size. Extraction efficiency in % for SAE fine dust. SAE fine $\geq$99.5 %. High probability of

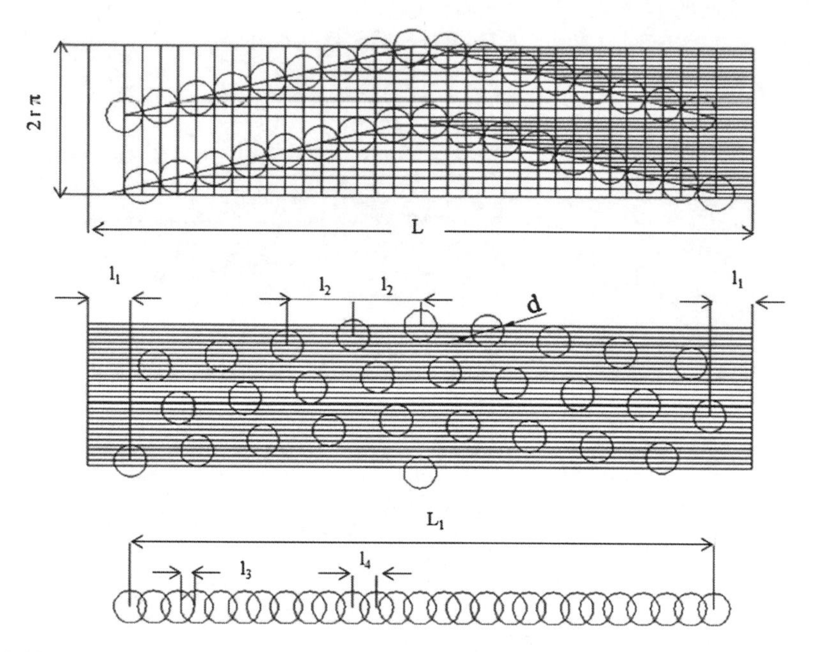

Fig. 3.15 Design of rotor helix (MV-4)

extraction found even 1 μm particles, and the virtually certain extraction of particles larger than 3 μm in size.

Another important requirement for dry air cleaners is, in addition to specified efficiencies, a high dust-holding capacity. Service life can be defined in terms of the quantity of dust taken up by the filter with a given increase in air-flow resistance. In order to ensure a high dust holding capacity, the paper folds are provided with surface contours that maintain clearance, guarantee an unhindered flow of dust-laden air into the folds and simultaneously prevent the folds from contacting each other.

3.2.4 Flail System and Soil Digging Profile

The flail head assembly consists of the protective shield, which sits over the flail head assembly to block the blast of the mines away from the machine, protective bar, mine clearing tool shaft, chain & hammer assembly, supporting wheels, Fig. 3.14.

Flails are attached to the rotor along the helix line, produced by coiling the "triangle around cylinder", Fig. 3.15. Two helixes are set up on rotor, with phase shift of 180°. Between the flails, lag is 21°. It is important to determine minimal number of hammers grasping the soil in order to provide required strike density. Soil digging hammer is usually of "mushroom" shape, which is suitable for soil cutting and crunching, and often the change of hammer shape is not required. For safe digging and destroying of AP mines down to the depth of 20 cm it is required to have 75 kW per meter length of the rotor.

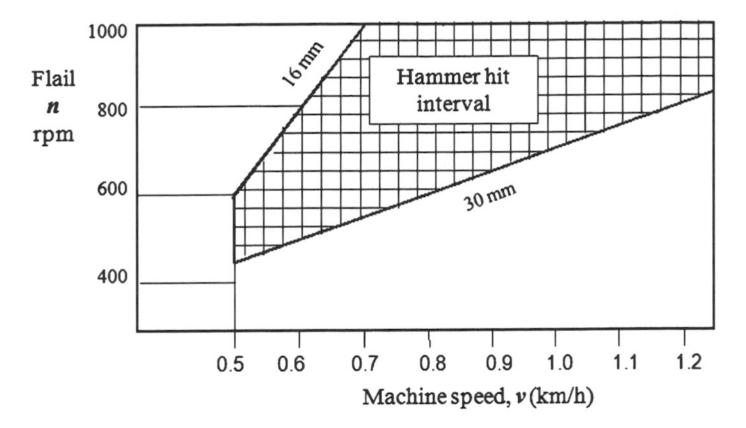

Fig. 3.16 Diagram of soil digging density, machine speed—flail rpm

3.2.4.1 Soil Digging Density

In the example of neutralizing the smallest AP mine type PMA-2, it is shown that the smallest and the largest feed of hammer shear (for soil digging) can be determined. These are very strict requirements for clearance density (the mine fuse of 16 mm in length), achieved by hammer shear of 16 mm, and machine forward speed of less than 0.75 km/h and greater rotor rpm. Higher machine forward speed is achieved with hammer shear of 30 mm and lesser rotor rpm, which provides greater efficiency. Hammer feed 30 mm is cca. 50 % of the smallest AP mine PMA-2 diameter (Ø68 mm).

An important issue is the density of hammer impacts against the soil in order to ensure safe destruction and elimination of mines. Soil processing density is determined by the vehicle movement speed and the flail rotation speed. In order to use the machine it is necessary to harmonize the machine movement speed v and the number of flail rpm n, i.e. to determine the cutting feed of the impact tool S. This distance between the points where hammer hits the ground, has to be less than the size of the smallest mine, i.e. of the critical value "c" (16–30 mm). The machine has to maintain required density—hammer hit interval, according to the equation:

Machine speed

$$v = Sn \text{ (km/h)} \tag{3.9}$$

Flail rpm

$$n = v/S \tag{3.10}$$

S hammer shear (feed)

given by the diagram of soil digging density, Fig. 3.16.

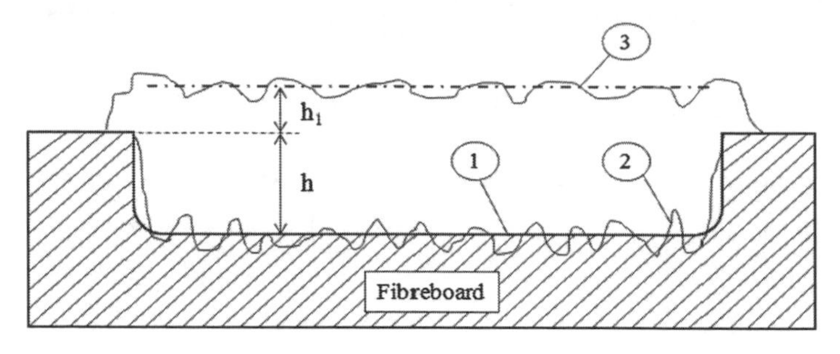

Fig. 3.17 Soil digging depth profile on fibreboard *1* ideal profile of curved edges of flail cut, *2* real profile of curved edges of flail cut, *3* loose layer of soil surface, *h* soil digging depth, h_1 loose layer of soil height

3.2.4.2 Soil Digging Depth Profile

The hardness of the ground plays a major part in determining the ability of a machine to penetrate to a given depth. In very hard ground the machine will be less capable and in softer ground the machine will be more capable. If there are soft spots or soft zones in a section of hard ground, the machine will be able to penetrate better in those soft spots/zones than in the surrounding hard ground. In order to measure the ability of the machine in the hard ground, it is important not to compromise the data by creating soft spots/zones in the area where the measurements are to be taken. It is possible that there is untreated soil that may hold hidden mine danger. Therefore the machines are tested in several soil categories and several digging depths, 10 cm, 15 cm and 20 cm. Soil dogging depth profile are measured using fibreboards in topsoil. Soil digging depth profile using flail is given in Fig. 3.17. Quality of mine clearance is ensured according to realistic digging profile. In doing so, soil digging depth (h) and loose layer of soil height (h_1) has to be distinguished. Movement of a demining machine on such loose layer with weak adhesion level sometimes presents a problem of its usage.

3.2.5 Machine Control

For a light machine that will operate on mine suspicious areas, it was necessary to develop a remote control system, consisting of a *microprocessor box and the control console.* On the machine—in the box itself, is a variable power transmitter, receiver for feedback information and battery. Several synchronized and highly-integrated processors are monitoring the whole operation of controlling the machine and diagnostics. Using the light control console, carried by the operator, one can directly remotely control the machine from a safe distance or from an accompanying armoured vehicle.

Fig. 3.18 Operator control unit (OCU) with micro operator hand-held unit (OHU)

The operator remote kit consists of the operator control unit (OCU) and the operator hand-held unit (OHU), Fig. 3.18. The OCU has a digital read out screen, which displays the machine's current status and functions performed. It incorporates micro processors that can halt all performed functions in order to prevent damage to the machine and the assemblies control operation of machine. Inside the box is a transmitter of variable power output, receiver for feedback information and a battery. A self-testing system warns the machine operator about possible malfunctions by means of tone signal on the OCU.

Remote control of the machine is accomplished by using the OHU. Ergonomic design and lightweight enable ease of use. For reading status of the buttons and sticks there is a processor built into the unit. The same processor monitors communications between the OCU and the OHU itself. The communication between them is serial and bi-directional. The OCU reads the data coming from the OHU with the approximate frequency of 10.000 times per second, which ensures a quick response and reduction of the possibility of incorrect information, in case of any external interference or EMI that could influence the accuracy of data.

3.2.5.1 Electronic System

The operator control unit contains the rechargeable battery, 2 × 20 LCD, 5-button keypad, 920 MHz integrated spread spectrum transceiver, Fig. 3.19. Attached to the operator control unit is the operator hand held unit. This is an ergonomically designed game controller shielded to prevent interference from external sources. There are 15 discrete buttons and two thumb joysticks housed in the controller. Included in the design is a permanently mounted environmental compliant connector that is used to charge the internal battery of the operator control unit (Figs. 3.20, 3.21).

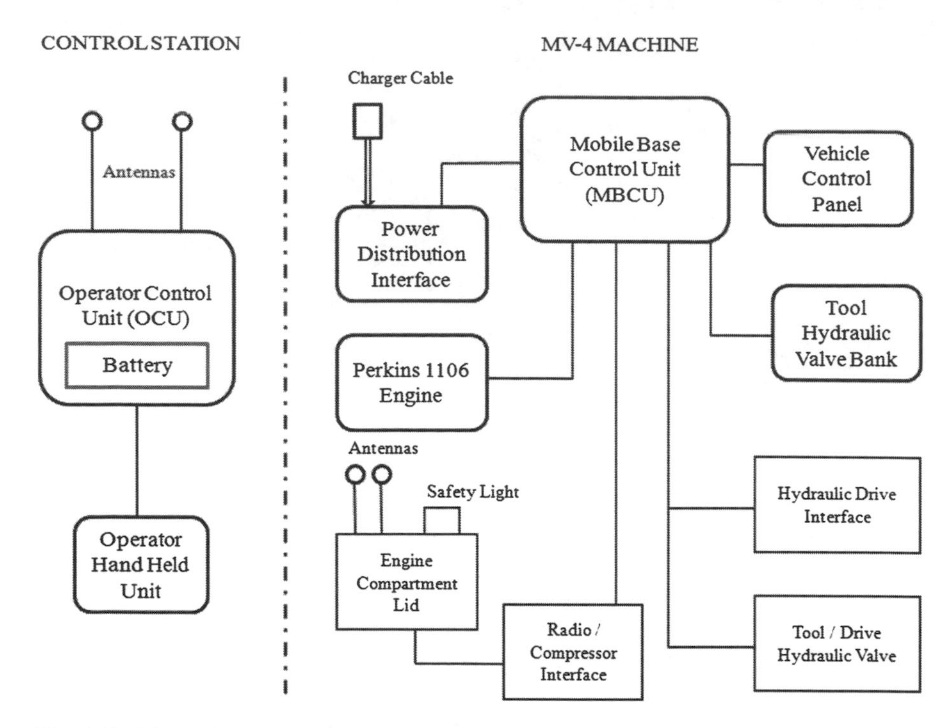

Fig. 3.19 Electronic system scheme electronic system scheme

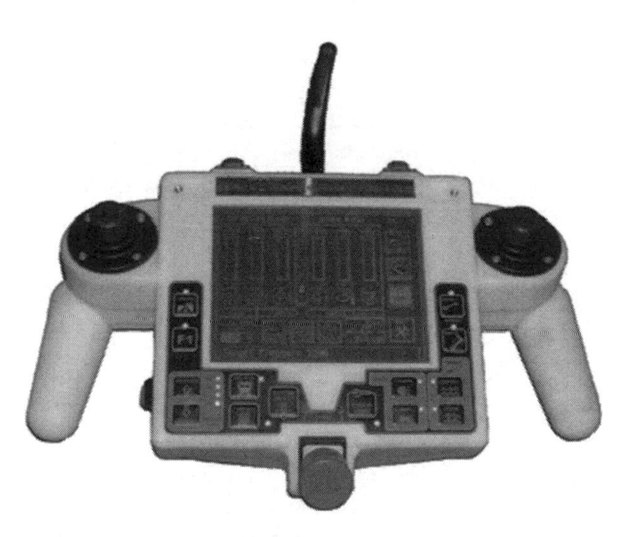

Fig. 3.20 Macro operator hand-held unit (Reproduced with permission from Ref. [10]) *6 joystick axes, Hi Res Touch screen display, Customizable keypad with LEDs, Customizable handles, Built in batteries, Built in semi duplex transceiver, Color TFT display ready, Ability to flash MBCU & VCP firmware, 8Mbyte telemetry storage, USB & Serial interface to PC, E-Stop, EMI protected, ESD resistant, Battery life—8 h normal, 16 h power save and 1.1 kg mass*

Fig. 3.21 Video operator hand-held unit (Reproduced with permission from Ref. [10]) *4 joystick axes, 2 × 16 char display, colour video TFT, OSD, external NiMh or LiPo battery, built in transceiver, built in video receiver, serial interface to PC, EMI protected, ESD resistant, battery life—8 h, 5 kg mass*

Wireless controller specifications

Vehicle side

Durability:

- Water-resistance, designed for outdoor usage
- Weatherproof connectors
- Heavy duty mounting bracket
- Shock resistant, designed specifically for mobile use
- 6–8 g—force rated shock resistant
- EMI shielded (tested at 50 Vpm)
- ESD 15 000 V discharge
- Temperature range −30 to 75 °C

Operator side

Durability:

- Water-resistance, designed for outdoor usage
- Weatherproof connectors ABS plastic enclosure
- 2 g—force rated shock resistant
- EMI shielded (tested at 50 Vpm)
- ESD 15 000 V discharge
- Temperature range −20 to 55 °C

3.2.6 Machine Characteristics

The light demining machine is designed to have a low profile and aerodynamic hull shape facing the explosion front, which is a completely authentic approach to the mine clearance vehicle development. The machine has been developed according to real field requirements and mine detonation, and not by adapting an existing working machine loader or a forest machine. Low specific ground pressure is achieved by the use of tracks. Hull is made of amour plates of 8 mm thickness. Impact hammers are made of steel highly resistant to wear-out and explosion detonations. Chains and hammers can be replaced quickly if damaged. Safe machine operation requires only the power engine drive which provides the flail with power of 75–100 kW per meter length of the rotor. On wet ground, sand, and stone, the supporting wheels or flail roller is replaced by sledge type support system.

Technical characteristics of the light demining machine

Machine type	MV-4
Machine mass	5.000 kg
Machine dimensions	2.60 × 2015 × 1.45 m (L × B × H)
Machine dimensions with tools	4.45 m × 2015 m × 1.45 m
Engine, Perkins	129 kW/2200 min^{-1}, water cooled
	695 Nm at 1400 rpm
Transmission	Hydrostatic, regulated pumps
Ground/MMP pressure	46 kPa/125 kPa, track width 300 mm
Operating tool	Flail/tiller/roller
Flail rpm	0–900 min^{-1}
Flail diameter	900 mm
Machine operating speed	0.5–2.0 km/h
Ground clearing density	16–30 mm
Maximum speed	5 km/h
Clearing depth and width	10/20 cm; 1.8 m
Number of flail hammers	34/40 pcs
Hammer mass	0.6–1.0 kg
Hammer service life-time	20–50 wh
Fuel, oil tank	70 l; 200 l oil cooler
Machine protection, guard	*Armox*/*Hardox* 400/8 mm
Operator distance	100–200 m, frequency 433 MHz
Efficiency	500–2000 m^2/h

Estimate of machine efficiency

Work capacity on the given testing area depends on the working conditions. The actual machine work capacity can be compared to the calculated work capacity. The calculated demining machine work capacity is the soil volume (V) that machine can bring up to the required clearance level over certain time (T), that is $U = V/T = B\ h\ L/T = B\ h\ v$ [m^3/h].

For practical reasons, for constant digging depth h = 20 cm, the work capacity is calculated in m^2/h or ha/h. $U = 3B\ v\ k_t/z$ [m^2/h]; *B—operating width, v—machine speed (L/T),* k_t*—factor of time use, z—number of runs (1 or 2).* Since the machine operating speed v depends on the soil hardness, work capacity will mostly depend on the machine operating conditions, in lighter, medium and heavy ground category, Fig. 3.22.

3.2.7 Assessment of Acceptability

For deminers and mechanization the most dangerous of all antipersonnel mines is the bouncing—fragmentation mine PROM-1. These mines were laid at several points and sideways from the edge of the working device for testing. After

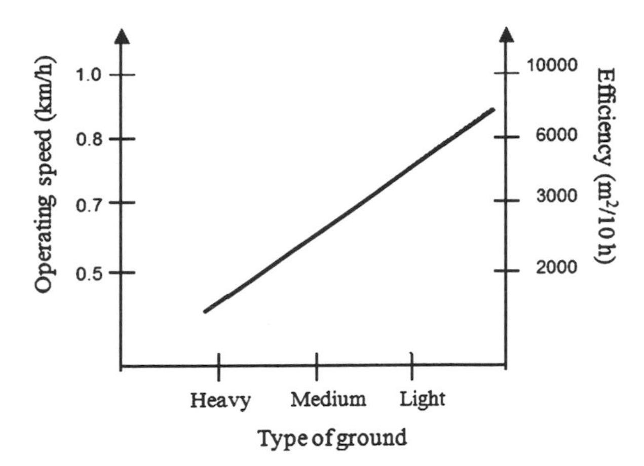

Fig. 3.22 Diagram of an estimate of machine efficiency

detonation, small dentures from mine fragments were seen on the working machine, but the machine remained operational. Operator, located behind the machine at a distance of 50–200 m is protected from fragments by protective equipment or shielded by the armoured vehicle. The machine is not intended for anti-tank mine clearance. However, in practice on mine clearance projects, anti-tank mines were destroyed by the 5 ton machine with or without detonation. After detonation of anti-tank mines the working device of the flail may be damaged completely, but the driving mechanism remains in good working order. Then the working part of the flail needs to be replaced by a spare device. This means that the reparability criteria, i.e. machine self-sustainability against mine damage that could be expected in mine suspected areas, has been satisfied.

According to the testing plan and program, the attention is paid to machine testing in controlled conditions and on real minefields to antipersonnel mines and the survey of the land after mechanical treatment. Some testing results of the 5 ton demining machine are:

In controlled testing conditions, on the project near the city *Otocac*, 19 mines were used: PMA-1A (5 pieces), PMA-2 (5 pieces), PMA—3 (5 pieces), PMR-2A (2 pieces), PROM-1 (2 pieces). All PMA-1A mines were activated, that were buried at the depth of 10 and 20 cm. PMA-2 and PMA-3 mines, buried at depth of up to 15 cm were activated, and at the depth of 20 cm were crushed. These results are showing that antipersonnel mines, buried at the depth of 15 cm, would be activated, and at the depth of 15–20 cm crushed and destroyed. Antipersonnel pressure-activated mines are very rarely intentionally laid at the depth of 20 cm, but can be found at this depth due to different shifts in the ground during previous 10 or more years. Results also show that the flail machine featured a good arrangement of chains and excavators, that it reaches the depth of up to 20 cm and that antipersonnel blast-activated mines down to this depth will be destroyed. Out of 4 antipersonnel blast mines, 3 mines were activated and 1 mine was partly damaged. The resulting damage on the machine and the flail were insignificant for its operation. No corrective maintenance activities were necessary. Vital machine parts and components are very well protected.

In testing conditions on the minefield in eastern Croatia, 5 ton demining machine has achieved the following results:

- average machine efficiency 1,543 m^2/h,
- soil treating depth is around 20 cm,
- the soil was of 3rd category, with increased humidity level,
- vegetation was low grass (20–25 cm high) with occasional bushes.

Acceptability in cross-country mobility

Driving in slope of 60–70 %
Driving in side slope of 35 %
Mine clearing in slope of 35 % up and down
Mine clearing in side slope of 20 %
Mine clearing 25 cm to a wall
Cross-country capability (ditch, ford, obstacle)
Water fording 70 cm depth.

Conclusions

A light remotely controlled tracked armoured machine design has small dimensions and high cross-country capability. Operating tool for digging and mine clearance is a flail mounted at the front, which destroys the smallest antipersonnel mines and the most dangerous antipersonnel bouncing mines. Mine destruction is performed by the force of impact tools of appropriate shape—hammers at the end of the flail chains. Flail rotation and hammers strokes are digging the soil down to the depth of twenty centimetres, together with mine destruction or activation.

The machine is intended for mine clearance of fields, mine-suspected areas around houses and areas that cannot be accessed by bigger machines. The performed calculations of the machine have resulted in design solutions that showed good results in practice. Required force of the hammer impulse has been determined in order to overcome cutting resistance and soil density resistance.

Required speed of the machine and tool has been calculated, which determines the soil processing density and consequently the flail working tool has been designed. Hydrostatic transmission for vehicle movement and digging of ground has been developed. Low centre of gravity, low profile and aerodynamic hull shape facing the explosion front have been designed, which is a modern approach to demining machine design development. Machines tests results in real and

Fig. 3.23 Working tool is smooth disc rims ROLLER MVR-1, discs "adjusting" to the soil profile, option—ribbed disc rims (Reproduced with permission from Ref. [10])

controlled conditions are presented, showing high grade performance regarding mine clearance quality.

3.2.8 Light Demining Machine with Rollers

For demining of areas where AP mines were laid, particularly for demining of roads for movement of convoys through unsafe areas, use of discs for demining is realistic option. When demining with rollers, there is no dust, which would decrease road visibility. For example, on very dry terrains, use of flail or tiller causes big and dense dust clouds, which hampers machine demining. In these conditions, demining rollers (discs) that are pushed in front of the machine, can be used. Light machines have enough power for pushing discs sections. AP mines are buried close to surface, enabling mine activation by discs.

3.2.8.1 Rollers Design

Demining rollers device consist of disc section, shields and bearing frame, Fig. 3.23. Disc pressure on mine fuse is causing mine activation. AP mine explosion does not cause damage on discs or shields. If light machine accidentally hits AT mine, it can be thrown away, and discs could be damaged. To achieve better contact with mines and its activation, discs can shift vertically and adjust to the profile of rugged surface. That is why disc hole is larger than axis diameter for 200 mm. *Discs rims can be smooth or ribbed for cutting the soil and mine fuse activation, and to ensure discs rolling, i.e. to prevent disc sliding, which further causes mine pushing and not their destruction.* When demining minefield, working speed is 10–15 km/h. Rollers shortcomings are as follows: poor maneuverability and difficult change of direction, difficult use on slopes, poor mine clearing performance on soft soils. Reliability of mine activation using discs was explained previously in machine demining mechanics chapter.

Rollers device has to protect the vehicle on which is mounted, and probability that discs will hit the mine has to be higher than probability that tracks will hit the mine. This is achieved by width of disc sections being higher than width of tracks. It is assumed that the force of disc weight on the mine is always higher than force needed for mine activation, which is assured by disc weight safety factor.

Basic data:

- Basic demining machine—LDM
- Working tool—disc *ROLLER MVR-1*
- Disc section width 1,8 m
- Total disc section width 2,0 m
- Number of discs in Sect. 10 pcs
- Disc width 150 mm
- Disc weight 150 kg
- Total disc device mass 2 t.

3.2.8.2 Evaluation Factors

- Technical characteristics of rollers device
- AP mine destruction depth for different soil types and machine movement speed
- Working tool durability against AP mine activation
- Efficiency of machine remote control device
- Need for other survey and demining methods

Phases:

1. preparation of lanes on test range
2. lanes with different soil type and different AP pressure activated mines
3. machine testing on AP fragmentation mines
4. deminers survey of testing lanes

AP pressure activated mines (PMA-1A, PMA-2, PMA-3) were set up on two test lanes, one made of compact soil and the other made of compact sand, at distance of 5 m and at different depths (0.0 cm, 5 cm, 10 cm).

Neutralization of AP pressure activated mines using rollers device for demining the passages, at depth of 0 cm is 90–100 %. If depth increases to 5 cm and 10 cm—probability of mine destruction decreases.

Neutralization of AP pressure activated mines in compact soil is around 85 %, and in compact sand around 50 %. If soil load bearing capability decreases—probability of mine destruction decreases too. Detonation of AP pressure activated mines does not leave functional damages on discs or basic machine.

3.2.9 Mini Demining Bulldozer

In demining, mini bulldozers are displacing mine suspicious soil. They are also used, after explosions, for clearing the ruins, vegetation and obstacle removal, and for burring dikes and trenches. Dozer device is mounted on the front part of the machine, and consists of dozer mechanism, dozer knife and hydraulic installation. Machine is remotely controlled. Dozer knife is about 10 % wider on each side then the machine. On each side of the knife, wings can be mounted for directioning of dug soil. "U" shaped knife is the best for dozing the soil on larger distances. Shock absorbers can be mounted on the knife.

Data of dozer knife blade

- Soil cutting resistance:

$$R_1 = k_1 L h \, [\mathrm{N}] \tag{3.11}$$

k_1 cutting resistance
L cutting tool width/knife blade length
h digging depth

- Specific pushing force at knife blade:

$$q_h = Fv \,/\, L \, [\mathrm{N/mm}] \tag{3.12}$$

Fv traction force
Fv $\approx G\varphi$; φ—adhesion coefficient

- Specific vertical pressure to knife blade:

$$q_v = Fy \,/\, A \, \left[\mathrm{N/mm^2}\right] \tag{3.13}$$

Fv force on the knife blade in vertical plane (due to knife pushing)
A knife blade area, $A = bL$
b knife blade width

Mini dozer MVD-1 is developed in accordance with requirements for maximal machine height of engine, Fig. 3.24. Machine has low CG and aerodynamic body shape facing the centre of explosion, providing that possible blast wave and mine fragments are passing above the machine. Machine has high cross-country mobility, due to very low specific track pressure to the soil. Because of its low profile, and remote controlled operation it is suitable for multi-purpose tasks and missions.

Fig. 3.24 Tracked mini dozer (*Source* Ref. [10])

Teeth can be mounted to the knife blade, for easier digging of hard soil. Teeth are made of materials resistant to wear-out, usually steel alloys (martensite CrSiMo steel), which are thermally treated. Alloying elements (Cr, Mo) are creating carbides, important to maintain tool hardness, exposed to high loads and abrasive wear-out. Steel hardness is around 500 Brinell. Teeth are replaceable.

Technical data

Dimension with tool, L × W × H : 3480 × 1600 × 830

Mass	4.5 t
High	830 mm
Engine	65 kW/2800 min^{-1}
Efficiency	60–100 t/h
Tool	Bulldozer blade
Maximum slope	35 %

3.2.10 Advanced Demining Systems

Clearing of mine suspicious area in urban areas stagnates behind development of other countermine equipment. For demining in urban environment with limited maneuvering space, for removal of scattered and bulk UXO, and for antiterrorist actions, mobile *machine demining* mechanization is required. That required development of demining mechanization for specific conditions. *Tele-demining machines*, based on tracked vehicle with telescopic arm and different tools is conceived, that will contribute to their dual use, Fig. 3.25.

Beside the need for that kind of equipment, commercial justification for development of such equipment is also important. Dual requirements are determined for development of that equipment, which will be used for demining and material handling purposes. A family of *tele-demining machines* was

Fig. 3.25 Tele-operated demining system (Reproduced with permission from Ref. [10])

conceived as possible solution for urban countermine and commercial tasks, so called dual use demining technology, Fig. 3.26.

3.2.10.1 Tele-demining Machines

Dual use demining technology considers:

1. use of advanced *tele-demining* technology in mine suspicious area in urban areas
2. use of advanced *tele-demining* technology for handling of hazardous materials

Progress is seen in modernization of light demining machines, which can easily fulfill stated requirements by the use of telescopic arm, which should be equipped with easily replaceable tool and required outreach in working area. For example, for managing hazardous materials and material manipulations in bases and depots and in contaminated conditions, a model of logistic countermine vehicle, so called *tele-demining forklift*, is conceived.

User requests

1. Demining of suspected areas—surface cleaning using flail
   - lowering the time required for demining of mine suspicious areas
   - demining in forests and civil engineering areas
   - variety of use of demining tool and removal of hazardous objects
2. Restacking of special cargo
   - safe loading and unloading of hazardous cargo
   - restacking of cargo inside the base, depot, or in the field
   - variety of use of tools, gripers, shovels…

Advanced *tele-demining* machine (TDM) development is based on:

1. Chassis of light demining tracked or wheeled machine
2. Variety of tools according to cargo type: forks, shovels, gripers, crane, etc.

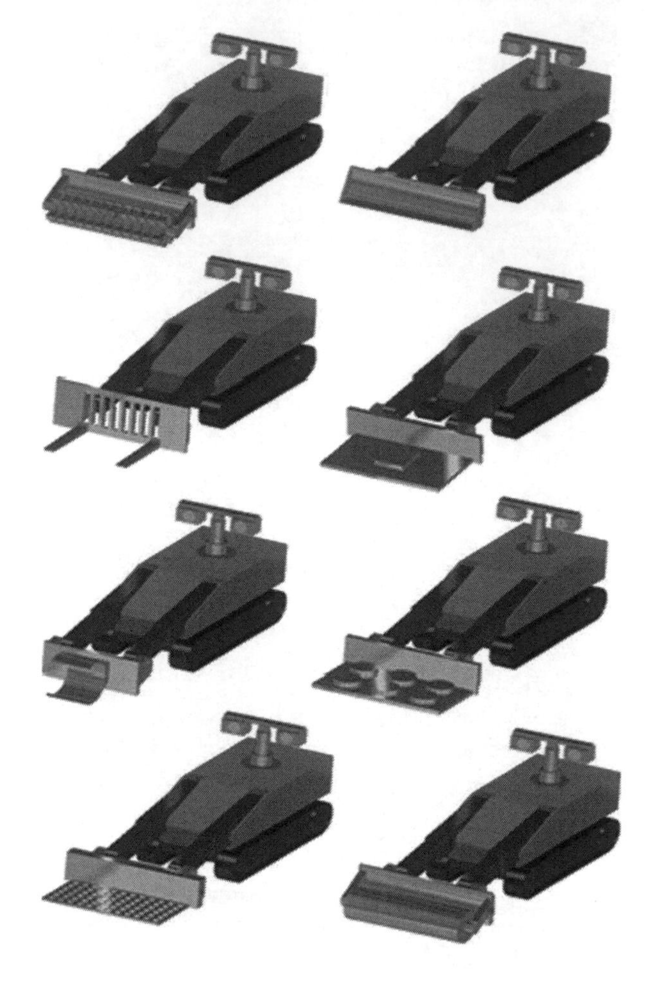

Fig. 3.26 *Tele-demining* machine family; mission essential module, exchangeable tools: flail, loading bucket, forklift *teletruck*, magnetic plate, gripper, multi-sensor plate, detonating grid, roller

3. Technology for manipulation of hazardous material (higher efficiency, lower costs)

   Goal: to develop an advanced *tele-demining* machine family for dual use and payload of 3 tons:

1. TDM for detection—detection vehicle
2. TDM for humanitarian demining—flail
3. TDM with demining rollers
4. TDM with detonating grid
5. TDM with mini-loader, dozer
6. TDM with magnetic plate
7. TDM with multi-sensor plate
8. TDM with clamps/grippers for barrels or other objects
9. TDM for stacking—*tele-forklift*

Basic requirements

Use in different demining conditions: in limited space, above obstacles, out of depots, in the field.

Use of different tool types: flail, fork, grab, clamps. Easy replacement of tools.

Adjustable tool outreach and height within working area (5 m).

Tool girder twist

(110°).TDM payload: 3 t, outreach 2–5 m.

Flail-additional tool.

Mass, 5 t.

Speed,

20 km/h.

- Development project of *tele-demining* machine includes:
  - platform, based on light demining machine chassis
  - telescopic arm superstructure
  - design of hydraulic system for cargo lifting
  - stability monitoring system development (software)
  - tools development: flail, rollers, forks, shovels, gripers, grids,..
  - integration of multi-sensors intelligent systems demining

Basic machine

Light demining machine is selected for *tele-demining platform*, due to dimensions, and suitability for the use in areas that are restricting mobility, such as blocked city roads, parks, yards, dikes or trenches. Chassis itself, due to its 5 t weight and light specific pressure to the ground, is suitable for use on all terrain types. Machine is powered by Diesel engine with hydrostatic drive, which provides good mobility characteristics. It can overcome the slope of 60° and side slope of 35°, and water-fording depth is 45 cm without any extra preparations. Due to its small dimensions and drive characteristics, 360° turn in place is possible. Additionally, it can cross dikes of 30 cm depth and width of 50 cm, or vertical obstacle height of 30 cm. Light demining machine is logistically acceptable, because it is transportable by 5 t trucks, helicopters, aircrafts or standard 20′ container.

3.2.10.2 Tele-demining Dobotics

Development of first robots used in military operations originates from the WWII, when engineering vehicle was developed for cutting the barbed wires. This "robot" was B1—remotelly controlled machine for minefields clearing, developed in 1939. Modern robots and innovative platforms are autonomous robots with manipulative features. Research and development of intelligent robots, for counterterrorist actions, military and other purposes, is intensified.

Research project goal

Develop a family of light autonomous machines (LAM) with a telescopic arm of dual use:

1. LAM with a demining flail
2. LAM with demining rollers
3. LAM with a multi-sensor plate
4. LAM with a magnetic plate/decoy
5. LAM with a loading bucket/dozer blade
6. LAM with apparatus for drilling in the mining industry
7. LAM with gripper
8. LAM for re-arranging freight—forklift.

The project encompasses:

- autonomous system for mechanical demining, autonomy refers to energetics, guiding system, controlling system, acquisition and documentary system, and guiding system by vehicle formation,
- modular system, designing the generic modular vehicle for purpose of many operations in demining, ability to apply in related activities, for reconnaissance, military support, for protection and other,
- effectiveness and productivity, logistic support.

Platform of technological development

5 ton demining machines is a good mobile unmanned platform for the LAM project exactly because of its dimensions, because it is appropriate for use in areas of limited mobility, like the jammed city streets, parks, yards and canals are. The tracked undercarriage has low mean maximum soil pressure and NATO mobility index, so it is very suitable for use on soft ground. The Diesel engine powers the machine, and the transmission is hydrostatic. Diesel-electric drive is a developing option.

3.2.11 Autonomous Unmanned Mine Clearing Robot Concept

The final goal of the proposed dual-use machine concept is the development of a fully autonomous unmanned mine clearing robot—AUMCR [4]. Autonomous operation of the described light-weight machine enables many operative and tactical missions to be performed in a safe, accurate, relatively fast and a cost-effective way. At present, research efforts are constantly being invested throughout the world in the development of various types of robots, both for military and humanitarian mine clearance. However, the majority of such solutions still offer rather limited levels of autonomy (for example, automatic obstacle avoidance systems) and are still heavily depended on a human operator remote control. On the other hand, there are only a limited number of sophisticated state-of-the-art robots, developed mainly for difficult military and anti-terror tasks that feature fully autonomous operation.

Objective is to develop a modular, open-system platform that would enable the implementation of an advanced control system for the proposed AUMCR concept.

In such modularly structured system, various functions could be applied, ranging from simple semi-autonomous inspection of smaller mine-contaminated areas to a fully autonomous robot, capable of performing highly complex navigational and tactical missions. Moreover, cooperative and coordinated work of more than one machine in a mine field also becomes feasible and highly justified. The simplified structure of the AUMCR control system is shown in Fig. 3.27.

The low-level part of the system includes basic vehicle steering and maneuvering control, as well as other safety–critical systems (such as environment reconnaissance and vehicle localization, based on GPS technology). The controller onboard vehicle is interfaced to a human operator or a higher-level control system via wireless link. The mid-level controller does trajectory planning (mainly in 2D space) and also manages the transitions between different operating modes of the robot. Clearly, these transitions should be smooth (e.g. bumpless) and must not compromise the stability and safety of the robot, especially when travelling at higher speeds and/or performing dangerous operations with a working tool.

The highest level of the proposed AUMCR system deals with more complex and abstract tasks such as navigational or tactical mission planning. At this level, coordinated control of more than one vehicle in a minefield is achievable (i.e. robot formation guidance system). Except the main three levels of control, the AUMCR system must also contain FDIA (fault detection, isolation and accommodation), as well as some reconfiguration parts, that are inevitable in an autonomous mode of any control system.

A very important issue in the proposed AUMCR design is its real-time mine detection system. At present, there are many commercially available sensing technologies with different levels of maturity, each having its advantages and drawbacks. The most widely used tools are based on metal detectors, ground penetrating radars (GPR), infrared (thermography) systems, various types of neutron energy bombardments, acoustic detectors (land sonar), other types of electromagnetic spectrum energy bombardment, and explosive vapour detection (electronic noses). The later sensors (usually implemented in MEMS technology) show very promising results, especially for detection of small, deep-buried mines with low-metal content.

An innovative approach to a more reliable mine detection system relies on combining the data from multiple sensors and different sensing modalities. The method is well-known as a multi-sensor data fusion, and is recognized as a tool that shows a big potential for significant increase of the detection accuracy and a lower false alarm rate. The data processing algorithm is usually based on Bayesian method of partial probabilities or the neural network method.

A proposed concept of the AUMCR vehicle, equipped with proper multi-sensor, multi-modality mine detection system, gives a solid base for a whole range of minefield survey applications. For example, one type of a single AUMCR (or multiple robots guided in a formation) could be used for mine detection, identification and marking its location in an electronic terrain map. After the entire

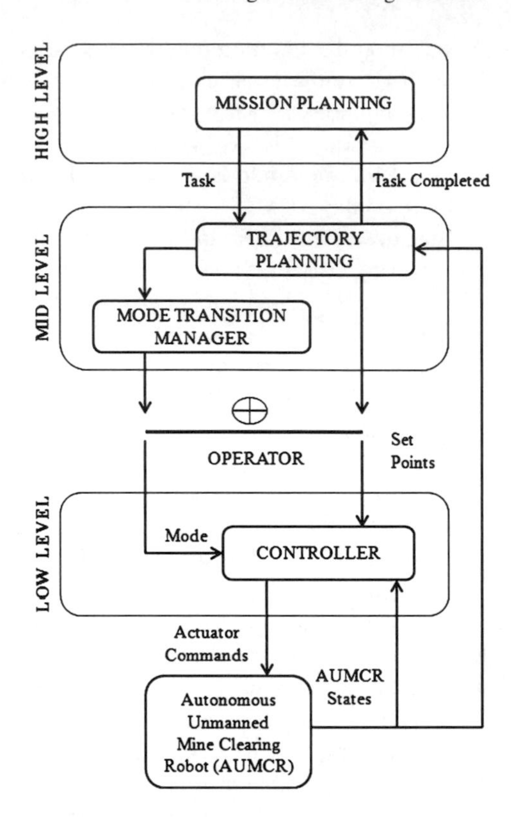

Fig. 3.27 Hierarchy of the AUMCR control system

minefield is thoroughly scanned, another robot with a proper working tool could be sent into the area with a dedicated mission of mine clearance. In general, for these types of tasks a semi-autonomous operation with human operator control is usually the preferable control strategy.

Conclusion

Special attention, when development of future countermine equipment is concerned, is given to equipment for mine threat neutralization in urban and forest areas. Basic characteristics of urban demining are:

- Mobility—problem is emphasized due to mine threats, narrow roads, concrete or wired obstacles, mines, ruins. Each crossroad is potential obstacle.
- Difficult to determine the exact position of mines and other obstacles in parks, municipal infrastructure, buildings, etc.
- Limitations to technical logistics resources—countermine equipment.

Technical development is focused on detection equipment and removal of possible threats. Development of remotely controlled machines and intelligent robots enables their application during peacetime, especially development of autonomous machines. *Dual use* application increases the countermine actions in different humanitarian operations. Tool replacement enables application of *tele-demining* machine as standard excavator, forklift or vehicle for manipulation of hazardous material. Its commercial side is highly appreciated, because instead of the use of several different machine types only one basic machine with different tools and sensors can be used.

Light autonomous machines have a market potential of dual use:

1. Their utilization can provide demining mine suspected areas in urban and forestry conditions, and at various anti-terror missions.
2. They provide economic benefits (hard mining works, relocation of freight in warehouses, loading–unloading of special freight, and other). Such machines of dual use constitute to justified investing in their development.

3.3 Design of Medium Demining Machine

Medium category of demining machines covers the range of 5–20 t. In the field of mechanical demining there is great potential for medium category machines, their development, production and role in demining operations, based on facts gathered during practical use of these machines. Because change in soil hardness during digging is very often, machine has sufficient excessive power available to overcome unexpected resistance. Medium demining machines reliable destroys AP and AT mines. Experience has shown that medium demining machines are very suitable for real demining conditions, i.e. they are compliant with realistic volume of demining projects, have required cross-country mobility and logistic support, and can be easily transported over long distances [5]. Use of modern demining medium category machines increases efficiency and safety in demining projects, and reduce demining cost.

Equipping of demining companies is based on all three demining machine categories. Light demining machines, with clearance width of around 2 m, are highly efficient for demining of smaller areas, terrains of different configurations and for vegetation cutting. Heavy demining machine, with clearance width around 3.5 m, are providing better efficiency when demining large and open areas, on different soil and vegetation types. Comparison of light and heavy machine characteristics is given in Table 3.1. There is significant difference in tool width ($\approx$1.5 m) for those two machine categories, and in size of treated area where light and medium machines are not efficient. This area size could be called as medium size project. Based on increase in efficiency, need for adequate medium machines of 15 t in weight and clearing width of 2.5 m emerged. Thereby, medium machines can be equipped for dual mode operation—operator manual control or remote control. Working tool for soil digging and mine neutralization can be flail or tiller, or their combination.

Table 3.1 Comparison of light and heavy machine characteristics

Advantages	Shortcomings
Light machines	
• Efficient in smaller projects • Low fuel consumption • Simple maintenance • Simple transport • Low demining costs	• Less spare power • Limitations considering soil type and vegetation • Low durability against AT mines
Heavy machines	
• Efficient in wide open areas • Good protection and resistance to explosive ordnance effects • Powerful engine, sufficient spare power • Able to achieve full digging depth on all soil types	• Reduced maneuvering capability • Difficult and complex transport at long distances • Increased fuel consumption • Higher procurement costs • Complex maintenance and operation

3.3.1 Use of Medium Machines

Use of machines in detection projects at deminer survey:

- Machine enters into surveyed area
- Machine treats whole area
- Machine destroys detected mines
- After machine, at least 15 % of deminer survey is performed using other method
- If an area is found to be systematically mine-polluted, area is considered as demining area category.

Use of medium machines in demining project:

- machine treats whole mine-polluted area
- machine destroys detected mines by activation, crushing or milling
- after machine, it is mandatory to use other method of deminer survey (manual method, dogs).

The goal is to develop medium category demining machine with working efficiency around 1500 m^2/h, which will efficiently perform demining in medium size projects and maintain good characteristics of both light and heavy machines. Demining machine can be developed on top of a tracked, Fig. 3.28, or wheeled undercarriage, Fig. 3.29.

Project requirements for wheeled demining machine (*Samson*)

- machine mass, 10 t
- clearing width, 2.5 m
- neutralization of AP and AT mines
- dimension 7.6 (8.0) × 3.0 × 2.7 m
- Diesel engine power 225 kW (306 KS) at 2300 rpm

Fig. 3.28 Medium demining machine with flail, RMK-02 (*Source* Ref. [11])

Fig. 3.29 Wheeled demining machine, *Samson 300* VILPO (*Source* Ref. [12])

- maximal torque 1090 Nm at 1400 rpm
- flail rotor with 56 flails on 5 helixes
- medium 4 × 4 wheeled demining machine category
- specific power on flail rotor (cca. 100 kW/m)
- maximum flexibility (chassis with joints)
- hydrostatic drive, controlled by computer:
  - prevents the possibility of engine and mechanical parts overloading; provides minimal fuel consumption and maintenance costs
  - automates machine and flail operation and adjusts machine speed in relation to soil category
  - automatic regulation system prevents possible increase of machine speed in relation to optimal working speed
- mechanical mine explosion compensation system which enables flail lifting to the height of 1.3 m, and two hydraulic compensation systems that are preventing transfer of explosion energy to the chassis
- road transport at speed 25 km/h; for longer distances transported on truck

Fig. 3.30 Medium demining machine, flail + tiller, MV-10 (*Source* Ref. [10])

- operation controlled manually or automatically, from the cockpit or remotely; for automatic steering computer is used
- hydraulic system: two closed hydraulic circuits, 420 bar, working hydraulic circuit pressure, 175 bar, hydraulic oil tank, 100 *l*.

Synchronization of machine forward speed and rotor angular velocity:

1. Rotor rpm is constant (n_r): if machine forward speed increases, tool shear increases too;
2. Tool shear is constant (S_t): if machine forward speed increases, rotor rpm increases too.

3.3.2 Requirements for Development

Project requirements for tracked demining machine (MV-10, Fig. 3.30)

- machine mass 15 t
- flail width 2.5 m
- soil digging depth 10/20 cm
- option: one flail—one rotor
- option: two flails—two rotors
- option: combination of flail and tiller
- AP and AT mine destruction—100 %
- maximum speed 10 km/h
- slope 60 %
- Remote control
- tracked demining machine with hydrostatic drive
- specific power on flail shaft ≥100 kW per one meter of shaft length
- fuel tank for 10 h working autonomy
- hydraulic system—closed hydraulic system
- air filtrating—high filtrating quality
- reliability for 8–10 h operations per day, in high temperature and dust density
- requirements—Clearance Requirements (IMAS)

- active protection—automatic fire-extinguishing system
- passive protection—additional protection for cooling system, no damages for flail in case of wire coiling
- working hour cost: between the cost of light and heavy machine cost
- logistic parameters for key systems: MTBF, MTTR
- transport: truck trailer.

Machine and flail

- mechanical reduction gearbox that drives hydropumps of variable flow for flail operation and machine movement
- flail hydraulic device consist two hydropumps, two hydromotors and two reduction gearboxes—chain drive, connected to flail rotor
- version: independent rotor drives for each flail
- version: independent drives for flail rotor and tiller
- tracked machine maneuvering device has two independent hydrostatic closed circuit power transmitter, one circuit for each track
- planetary gearboxes built in between hydromotor and tracks.

Design requirements

- soil clearing width 2.5 m

– soil digging profile 200 mm, capacity 1500 m^2/h
– hammer weight 0.9–1.3 kg
- specific flail power 100 kW per one meter of shaft length

– option: two flails on two rotors
– option: combination of flail and tiller
– Diesel engine
– fuel tank 450 l, for 10 h working autonomy
- hydraulic system—hydrostatic closed hydraulic system
- air filter—double filters—cyclonic and fine,
- active protection—built in automatic fire-extinguishing system
- passive protection—protective plates, made of ARMOX

– chain canvas, for protection of machine body
– additional cooler protection
– device for rotor protection against damage in case of wire coiling
- efficiency—production and field testing
- remote control—automatic selection of working frequency

Note: Machine dimensions, particularly width and length, are limited by machine transportability, by requirement for machine operation between buildings—machine dimensions should not exceed these parameters. Comparison of medium demining machines characteristics is given in Table 3.2.

Table 3.2 Comparison of medium demining machines characteristics

Features	MV-10 DOK-ING Ltd	RM-KA-02 D.D. Special vehicles Inc	SAMSON 300 VILPO Ltd	BOZENA 5 WAY Industry Ltd
Mass (t)	15	14	10	14
Drive	Tracked	Wheeled	Wheeled	Wheeled/tracked
Engine power (kW)	400–600	168	212	175
Length × width × height (m)	6.7 × 2.8 × 1.9	5.2 × 2.5 × 1.8	7.7 × 3.0 × 2.7	7.3 × 2.8 × 2.07
Machine width without tool (m)	2.22	2.00	2.50	2.40
Working tool	2 flails or flail + tiller	Flail	Flail	Flail
Flail diameter (m)	1,2/1,4	0.9/1,0	1,3	1,4
Tool rpm (min^{-1})	to 900	to 600	to 900	to 500
Clearing width (m)	2.3/3.0	2.0/2.5	2.5	3.0
Digging depth, max (mm)	450	300	300	300
Speed (km/h)	10	5	25	10
Steering	Remotely	Remotely	Remotely From cockpit	Remotely

Working tool—flail

Demining working tool is flail and consist of rotor and flails connected to rotor (chain + hammer). Flail material should be resistant to wear-out when used on different soil types, as well as resistant to blast loadings (centre of cross-section has to be of high tensile strength). Due to these requirements, cemented steel 16MnCr5 can be used for hammer material. By cementing, hard surface layers are achieved (resistant to wear-out), and core stays of ferrite-perlite structure, which is resistant to dynamic and blast loadings (tough). Chain is exposed to wear-out, primarily on tensile stress, due to hammer centrifugal force. Chain material can be tempered steel C45. Steel is improved (tempering and normalizing) in order to achieve high yield and tensile strength, with high toughness and dynamic strength.

Behind the flail, protective drum is placed, which protects from uncontrolled throwing of dirt and mine remains. Behind protective drum, on overall flail length, roller is mounted. On flail rotor, 50–60 flails can be mounted attached at required density in order to ensure that smallest AP mine would not be missed.

Synchronization of machine speed and flail rpm

Problem of neutralization of the smallest AP mine should be solved, i.e. to determine optimal ratio between machine movement speed and rotor rpm, as well as optimal positioning of flails on rotor. These parameters are adjusted so that mines of certain size are destroyed with no regard to the relative mine position in relation to the flail. Mine size is defined through critical mine dimension “c” (16–30 mm).

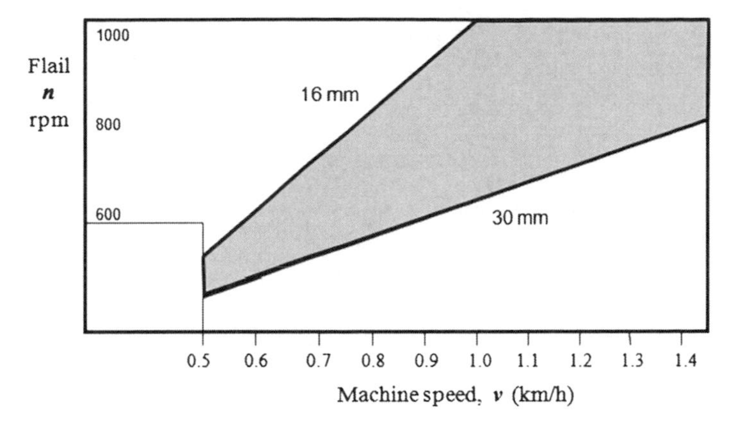

Fig. 3.31 Diagram of soil digging density, machine speed—flail rpm

Table 3.3 Machine working speed and flail rpm (rotor rpm)

Machine working speed	Flail rpm Hammer shear 16 mm	Flail rpm Hammer shear 30 mm
v (km/h)	n (min^{-1})	n (min^{-1})
0.5	521	278
0.6	625	333
0.7	729	389
0.8	833	444
0.9	938	500
1.0	1,042	556
1.1	1,146	611
1.2	1,250	667

According to the Eq. (3.9) $v = S\ n$, $n = v/S$, machine working speed (v) and flail rpm (n), for particular hammer shear (S) are given in Table 3.3 and Fig. 3.31. The best soil mine clearing quality is achieved by hammer shear of 16 mm, whereas machine speed is less than 1 km/h with greater flail rpm. Higher machine speed is achieved with hammer shear of 30 mm and lesser flail rpm, which provides greater efficiency.

3.3.3 Soil Resistance and Flail Power

(a) Soil cutting resistance

$$R_{ki} = z_n k_1 b S_t [N] \tag{3.14}$$

As z_n, k_1 and b are constant values; soil cutting resistance depends on shear S, digging depth h expressed in mm, and number of grasping hammers z_n, Table 3.4.

Highest cutting resistance R_k appears at digging depth of $h = 200$ mm, cut soil thickness—hammer shear $S_t = S = 30$ mm, and $z_n = 6$ grasping hammers (rotor length of 2.3 m). Due to occurrence of losses in soil cutting, cutting resistance has increased to 10 %.

Average values of specific resistance k_l, for soil category, approximately:

$k_l = 25$ kN/m^2—sandy clay, gravel
$k_l = 95$ kN/m^2—compact sandy clay, medium clay, soft coal
$k_l = 175$ kN/m^2—hard sandy clay with gravel, hard clay, conglomerate
$k_l = 320$ kN/m^2—medium slate, hard dry clay, chalk and soft plaster stone, marl

(b) Soil crushing resistance

Soil striking density for toll shear value of $S = 16$ mm is introduced, Table 3.5 [6]. Higher tool shear S, i.e. lower striking density, is causing high increase of soil cutting resistance. If soil is very hard, shear and crushing resistance should decrease in order to enable crushing of hard soils.

$$\begin{aligned} F_{\sigma 4} &= z_n b \sigma (S_{t1} + S_{t2}) \\ F_{\sigma 6} &= z_n b \sigma (S_{t1} + S_{t2} + S_{t3}) \end{aligned} \tag{3.15}$$

$F_{\sigma 4}$ total crushing force for soil crushing at depth, $h = 0.1$ m
$F_{\sigma 6}$ total crushing force for soil crushing at depth, $h = 0.2$ m
S_{ti} thickness of crushed/crunched soil layer
z_n number of grasping hammers
σ soil pressure strength (40×10^4–150×10^4 N/m^2)

(c) Flail force impulse, (Table 3.6) [6]

$$\begin{aligned} F_i t &= m_n v_0 \\ F_{iz} &= z_n F_i \;\; [N] \end{aligned} \tag{3.16}$$

m_h— hammer mass/1.2 kg
r —hammer rotation radius, $r = 0.55$ m
v_o —hammer circumferential velocity
t —layer cutting time
z_n —number of grasping hammers
F_{i4} —total impulse force for 4 grasping hammers at digging depth, $h = 0.1$ m
F_{i6} —total impulse force for 6 grasping hammers at digging depth, $h = 0.2$ m

Table 3.4 Cutting resistance

Cutting resistance [N]	Soil category k_1 [kN/m^2]	b = 60 mm S = 16 mm h = 100 mm z = 4	b = 60 mm S = 16 mm h = 200 mm z = 6	b = 60 mm S = 30 mm h = 100 mm z = 4	b = 60 mm S = 30 mm h = 200 mm z = 6
R_{k1}	25	43.22	74.50	81.03	139.68
R_{k2}	95	164.23	283.09	307.93	530.79
R_{k3}	175	302.53	521.48	567.23	977.78
R_{k4}	320	553.19	953.57	1,037.23	1,787.94

Table 3.5 Crushing resistance

Crushing resistance Hard soil category	b = 60 mm S = 16 mm h = 100 mm z = 4 $F_{\sigma 4}$ [N]	b = 60 mm S = 16 mm h = 200 mm z = 6 $F_{\sigma 6}$ [N]
1. Sand, acres	591.50	1,006.18
2. Sand, clay, acres	739.38	1,257.72
3. Clay of medium compactness	961.19	1,635.04
4. Compact clay, marl	1,330.88	2,263.90
5. Highly compact clay, marl	1,922.39	3,270.07

A diagram of resistance force and hammer impulse force in correlation with flail rpm is provided in Fig. 3.32. Soil cutting resistance and crushing force increase in correlation with soil category and hammer shear/feed, *S* (mm). Also shown is increase of impulse force with increase of flail rpm, resulting in better capability to overcome higher resistance when digging hard soils. The referent rotor rotation speed is set at 600 rpm. At this rotation speed even the highest soil crushing resistance can be mastered, such as compact clay or dry loess.

(d) Flail resistance moment

Resistance moment to flail rotation during operation includes static and dynamic resistance moments of rotating flail parts. Cyclic operation of each flail is assumed: hammer accelerates first and then nearly stops when crushing the soil. Hammer and chain inertia moment is approximated with spheres of equal weight, Fig. 3.33:

Rotor resistance moment

$$M_u = zJ\varepsilon + M_d \ [\text{Nm}] \tag{3.17}$$

- J — $J_h + J_c$ [kgm^2]
- M_d — moment of dragging force, $\geq 10\ \%\ M_u$ for $r \leq 500$ mm assumption
- J — total inertia moment of hammer J_h and chain J_c (kgm^2)
- z — number of grasping hammers
- ε — angular hammer velocity (s^{-1})

Data: $m_h = 1.2$ kg, $m_l = 0.75$ kg, $e_h = 0.55$ m, $e_c = 2.29$ m
$J_h = 0.3635$ kgm^2, $J_c = 0.0642$ kgm^2, $J = 0.4277$ kgm^2

Flail start up moment

$$M_p = pJ\varepsilon + J_r\varepsilon \ [\text{Nm}] \tag{3.18}$$

Table 3.6 Flail impulse force

rpm	Hammer circumferential velocity	Impulse force	h = 100 mm z = 4	h = 200 mm z = 6
n [min^{-1}]	v_O [m/s]	F_i [N]	F_{i4} [N]	F_{i6} [N]
400	23.04	184.31	737.23	1,105.84
500	28.80	287.98	1,151.92	1,727.88
600	34.56	414.69	1,658.76	2,488.14
700	40.32	564.44	2,257.76	3,386.64
800	46.08	737.23	2,948.91	4,423.36
900	51.84	933.05	3,732.21	5,598.32

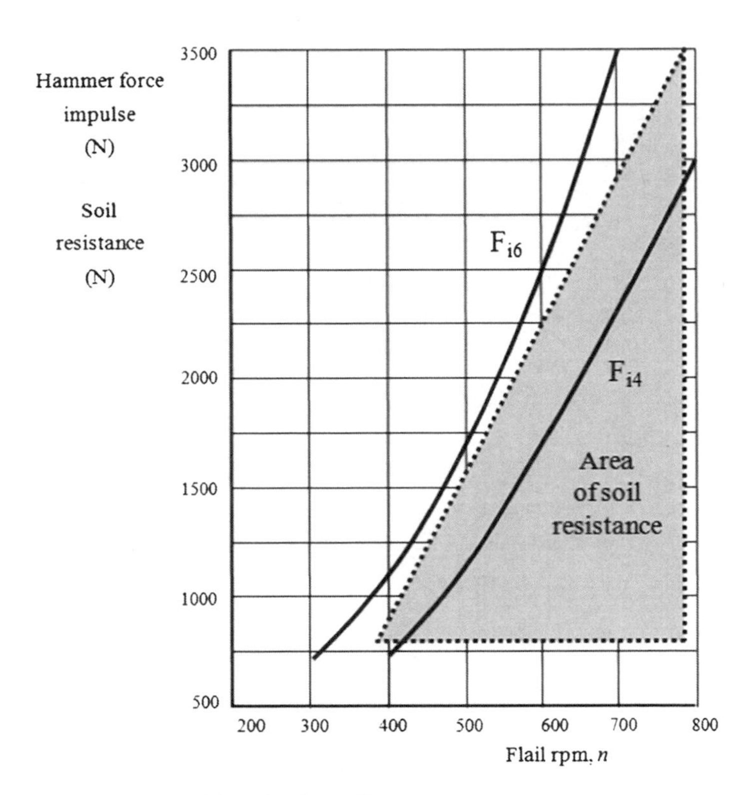

Fig. 3.32 Soil resistance and impulse force diagram

M_p moment required for start-up
M_{pi} start-up moment for various flail acceleration periods
J total hammer and chain inertia moment (kgm^2)
P total number of hammers
J_r rotor inertia moment (kgm^2)

$J_r =$ 0.516 kgm^2, $t_i = 0.5/1/2$ s.

Table 3.7 Digging moment and flail start-up moment

rpm	M_{ui} [Nm] digging moment		M_{pi} [Nm] start-up moment	
n (min^{-1})	$h_1 = 100$ mm $z = 4$ M_{u1}	$h_2 = 200$ mm $z = 6$ M_{u2}	$p = 46$ M_{p1} $t = 0.5$	$p = 46$ M_{p2} $t = 1$
400	525.52	788.28	1691.45	845.73
500	821.13	1,231.69	2,114.31	1,057.16
600	1,182.42	1,773.63	2,537.18	1,268.59
700	1,609.41	2,414.11	2,960.04	1,480.02
800	2,102.08	3,153.12	3,382.90	1,691.45
900	2,660.45	3,990.67	3,805.76	1,902.88

Table 3.8 Power required for soil digging power required for soil digging

n (min^{-1})	M_{u1} (Nm)	M_{u2} (Nm)	P_{u1} (kW)	P_{u2} (kW)
400	525.52	788.28	22.01	33.02
500	821.13	1,231.69	42.99	64.49
600	1,182.42	1,773.63	74.29	111.44
700	1,609.41	2,414.11	117.98	176.96
800	2,102.08	3,153.12	176.10	264.16
900	2,660.45	3,990.67	250.74	376.11

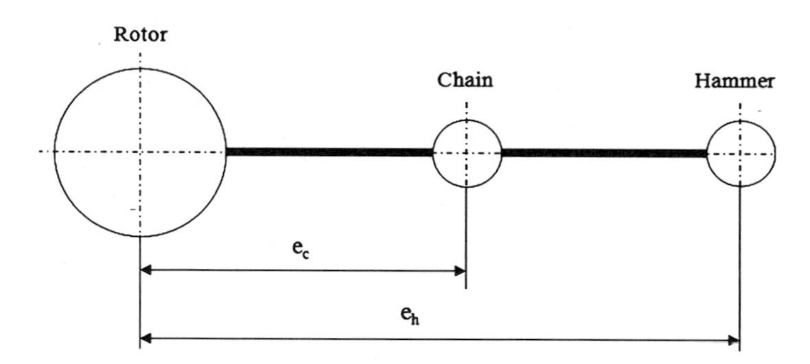

Fig. 3.33 Approximation of hammer and chain inertia moment

With increase in start-up time period, torque values required for flail free start-up M_{pi} decreases. Moment required for soil digging increases if flail rpm increases. For rotor rotation velocity of 600 rpm, hammer impulse force will efficiently treat all soil categories (Table 3.7) [6].

(e) *Power required for soil digging*, (Table 3.8)

$$P_{ui} = M_{ui}\omega_i \ [\mathrm{W}] \tag{3.19}$$

Table 3.9 Power at soil cutting, $S = 30$ mm

Digging depth	$h = 100$ mm				$h = 200$ mm			
Soil category	Cutting resistance R_i [N]	Flail path s_φ [m]	Flail work W [J]	Power P [kW]	Cutting resistance R_i [N]	Flail path s_φ [m]	Flail work W [J]	Power P [kW]
I	81.03	0.4	627.89	6.28	139.68	0.48	1,005.70	10.06
II	307.93	0.34	2,386.09	23.86	530.79	0.48	3,821.69	38.22
III	567.23	0.34	4,395.36	43.95	977.78	0.48	7,040.02	70.40
IV	1,037.23	0.34	8,037.31	80.37	1,787.00	0.48	12,866.40	128.66

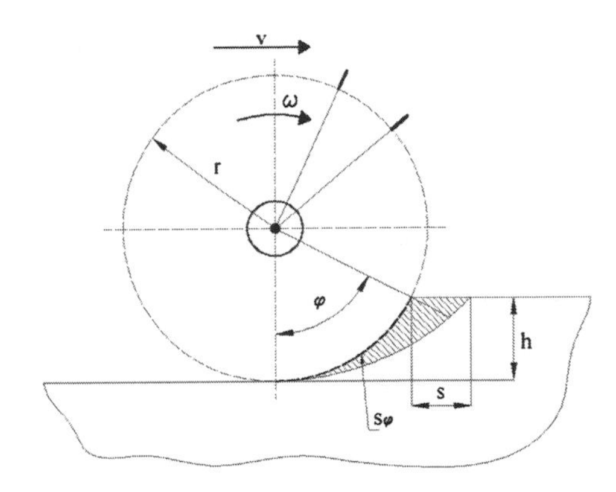

Fig. 3.34 Hammer path—circular arc

M_{ui} moment required for digging
ω_i rotor angular velocity.

(f) *Balance of power at soil digging*

Power at soil cutting

$$P = \frac{W}{t} = \frac{R_{ki} s_\varphi}{t} \text{ [W]}$$
$$R_{ki} = z_n k_1 b S_t \text{ [N]} \tag{3.20}$$

Hammer path (flail path) depends on digging depth h, so the required power is calculated for digging at depth of 100 mm and 200 mm, Table 3.9, (Fig. 3.34).

Hammer path is a circular arc (approx.):

$$S_\varphi = r\varphi$$
$$S_\varphi = \frac{2\pi r\varphi}{360} \text{ [m]} \tag{3.21}$$

$\varphi = arc\ cos\ ((r-h)/r)$, angle of hammer chain

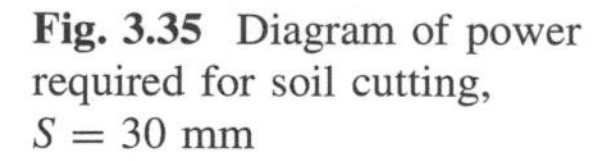
Fig. 3.35 Diagram of power required for soil cutting, $S = 30$ mm

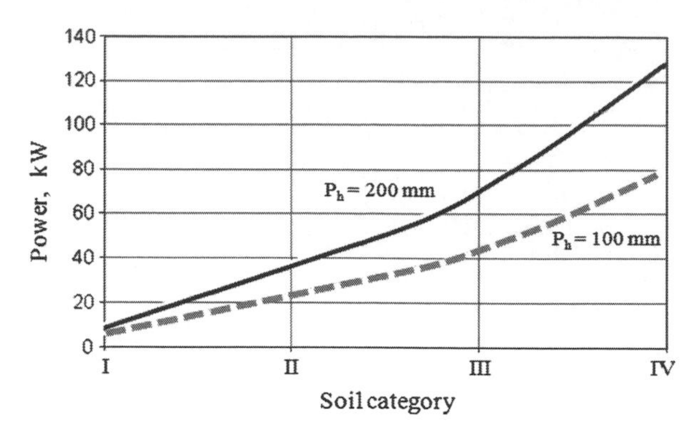

Table 3.10 Power at soil crushing, $S = 16$ mm

Digging depth	$h = 100$ mm				$h = 200$ mm			
Soil category	Crushing resistance R_i [N]	Flail path s_φ [m]	Flail work W [J]	Power P [kW]	Crushing resistance R_i [N]	Flail path s_φ [m]	Flail work W [J]	Power P [kW]
I.	591.50	0.34	4,583.42	45.83	1,006.18	0.48	7,313.34	73.13
II.	961.19	0.34	7,448.08	74.48	1,635.04	0.48	11,884.17	118.84
III.	1,330.88	0.34	10,312.74	103.13	2,263.90	0.48	16,454.99	164.55
IV.	1,922.39	0.34	14,896.24	148.96	3,270.07	0.48	23,768.26	237.68

Depending on soil category and digging depth, power consumption increases. For digging of soil category I and II, relatively small power is required, up to 40 kW, and for digging of soil category III and IV, increased power up to 125 kW is required, Fig. 3.35.

Power at soil crushing (Table 3.10)

$$P = \frac{W}{t} = \frac{R_\sigma s_\varphi}{t}\,[\mathrm{W}] \qquad (3.22)$$
$$R_\sigma = z_n \sigma b S_t\,[\mathrm{N}]$$

Power used for digging I soil category is 50–70 kW, and for digging IV soil category is 240 kW, Fig. 3.36. Comparison of the first and second power consumption diagram is shows that power required for soil crushing is a key factor for flail calculation and design.

(g) Flail specific power

Declaring the specific power per unit of flail shaft length is suitable for comparison and evaluation of demining machines. Specific power is engine power divided by rotor length:

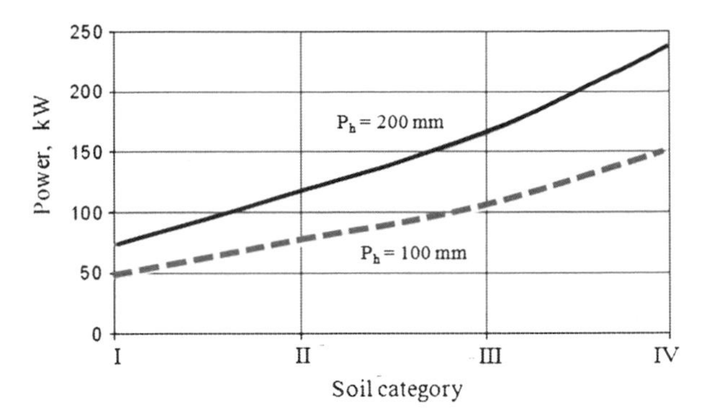

Fig. 3.36 Diagram of power required for soil crushing, $S = 16$ mm

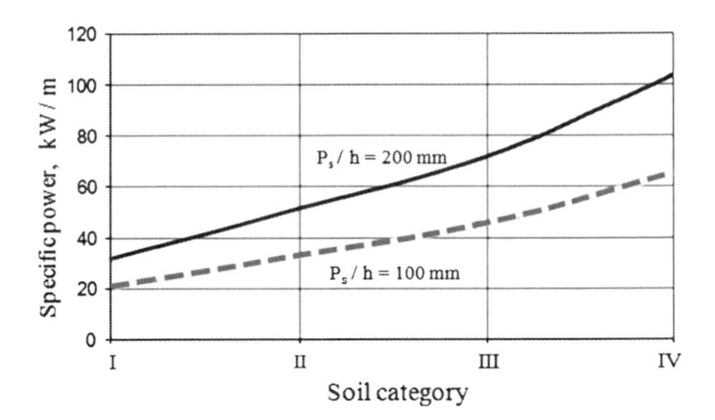

Fig. 3.37 Diagram of flail specific power

$$P_s = \frac{P}{L}\left[\frac{\text{kW}}{\text{m}}\right] \tag{3.23}$$

Specific power calculation confirms the assumption that for safe flail operation on hard soil a power of 100 kW per meter of rotor length is required. Diagram shows that with increase of digging depth from 100 mm to 200 mm, 50 % more power for soil digging is required, Fig. 3.37. Depending on soil category (I–IV), specific power required for safe flail operation on one-meter rotor length is between 30 and 105 kW/m.

3.3.4 Engine and Transmission

(a) Movement resistance and required power

Based on mobility requirements of tracked demining machine, drive power is determined. Movement resistances and tractive force are examined: rolling resistance, slope resistance and inertia resistance, reactive tractive force, Fig. 3.38.

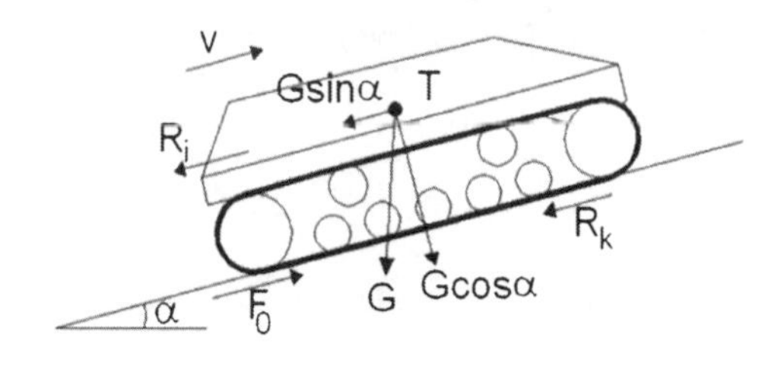

Fig. 3.38 System forces at machine movement

Movement resistance

$$\Sigma R = R_k + R_i + R_\alpha \tag{3.24}$$

$$R_k = Gf_k \cos\alpha$$
$$R_i = ma$$
$$R_\alpha = G\sin\alpha$$
$$\Sigma R = F_0$$
$$F_m = F_0$$

Track moment

$$M_{gi} = \frac{F_m r_g}{2}$$
$$P_{vi} = \sum Rv \tag{3.25}$$

ΣR movement resistance
R_k rolling resistance
R_i inertia resistance
R_α slope resistance
F_m drive traction force
F_o reactive tractive force
M_{gi} track moment
r_g track sprocket wheel radius
f_k tracks rolling resistance coefficient
P_{vi} power for machine movement
m machine mass
a machine acceleration

Calculation of movement resistance, required power and moment is given in the Table 3.11. It may be concluded that required track drive moment M_g and power P_v are increasing with increase of machine speed v and slope angle α.

Data: $m = 15\ 000$ kg; $\alpha_{max} = 60\ \%$; $r_g = 0.25$ m; $f_k = 0.15$;

1. Working speed: $v_I = 0.5$ km/h (1.0 km/h; 1.5 km/h)
2. Movement speed: $v_2 = 5$ km/h

Traction characteristics

Traction characteristics affect machine working efficiency. Traction characteristics are result of joint operation of engine and transmission. For increased resistance that cannot be balanced within working area of engine regulation characteristics, balancing is performed according to engine torque curve. However,

Table 3.11 Calculation of movement resistance, required power and moment

α (%)	R_k	R_i	R_α	ΣR	P_{vi}	M_{gi}
$v_I = 0.5$ km/h						
0	22,072.50	416.67	0.00	22,489.17	3,123.50	2,811.15
15	21,828.30	416.67	21,828.30	44,073.26	6,121.29	5,509.16
30	21,141.62	416.67	42,283.24	63,841.53	8,866.88	7,980.19
$v_2 = 5$ km/h						
α	R_k	R_i	R_α	ΣR	P_{vi}	M_{gi}
0	22,072.50	4,166.67	0.00	26,239.17	36,443.29	3,279.90
15	21,828.30	4,166.67	21,828.30	47,823.26	66,421.20	5,977.91
30	21,141.62	4,166.67	42,283.24	6,7591.53	93,877.12	8,448.94

this reduces engine rpm and machine technological movement speed. During that time technological operations are performed at unfavourable movement speed, which means that soil digging density was also reduced. Due to these facts, during machine design the application of additional engine power needs to be anticipated, so that excess power can be used to overcome increased resistance. For hydrostatic drive, regulation of flow and pressure provides power change.

Drive traction force: $F_m = \frac{M_{gi}}{r_g}$
Reactive traction force: $F_o = F_m$
Drawbar pull: $F_v = F_o - \Sigma R$

(b) Power of engine

Total engine power consists of power for machine movement and power for machine operation (Diesel Engine):

$$P_{DE} = P_r + P_v [\mathrm{W}] \tag{3.26}$$

$$P_r = \frac{P_{u1}}{\eta HS} [\mathrm{W}] \tag{3.27}$$

$$P_v = \frac{P_{vi}}{\eta PT \eta HS} [\mathrm{W}] \tag{3.28}$$

P_{DE} Diesel engine power
P_v power required for machine movement
P_r power required for machine operation
P_{ui} power required (Table 3.8)
P_{vi} power required (Table 3.11)
PT η efficiency coefficient of mechanical transmission (≈ 0.93)
HS η efficiency coefficient of hydrostatic transmission (≈ 0.7)
v machine working speed.

Machine working speed for required mine-clearing density is around 1 km/h, and machine should operate on slopes up to 20°. Efficient soil digging at depth of

Table 3.12 Selection of engine power

v (km/h)	α (%)	P_{DM} (kW)					
		n = 500 rpm h = 100 mm	n = 500 rpm h = 200 mm	n = 600 rpm h = 100 mm	n = 600 rpm h = 200 mm	n = 700 rpm h = 100 mm	n = 700 rpm h = 200 mm
0.5	0	66.21	96.93	110.93	164.00	173.34	257.60
	30	75.03	105.75	119.75	172.82	182.16	266.42
	60	81.69	112.41	126.41	179.48	188.82	273.08
0.7	0	68.18	98.90	112.90	165.97	175.31	259.57
	30	80.53	111.25	125.25	178.32	187.66	271.92
	60	89.85	120.57	134.57	187.64	196.98	281.24
0.9	0	70.18	100.89	114.89	167.96	177.31	261.56
	30	86.06	116.77	130.77	183.84	193.19	277.44
	60	98.04	128.76	142.76	195.83	205.17	289.43

h = 20 cm of all soil categories is achievable with selected optimal flail rotation speed of 600 rpm. Regarding all these parameters, a minimum required Diesel engine power P_{DM} = 183.84 kW (185 kW) is adopted, Table 3.12.

In order to provide engine running within an area of minimum fuel consumption (where the engine power is lower) and a greater power reserve to overcome additional movement resistance, the selection of Diesel motor is recommended with greater maximum power, e.g. that of 240 kW/2100 min^{-1} and engine torque of 1472 Nm/1400 min^{-1}. In case of using two working tools for soil digging, both flail and tiller, engine power can be twice as large.

(c) Main hydraulic components

According to hydrostatic calculation procedure, selection of main hydraulic components is as follows:

Drive	Hydropumps	Hydromotors
Machine movement drive	Volume: 28 cm^3 A4VG28EP2 D1	Volume: 56.1 cm^3 A2FM56
Flail operation drive	Volume: 90 cm^3 A4VG90EP2D1	Volume: 106.7 cm^3 A2FM107/61 W

3.3.5 Machine Efficiency

Machine efficiency depends on conditions in which machine operates. Calculated demining machine efficiency is soil volume V, with required soil clearance level, for certain time interval t:

$$U = V / t \left[\mathrm{m}^3/\mathrm{h}\right]$$
$$U = B\,h\,v \left[\mathrm{m}^3/\mathrm{h}\right] \tag{3.29}$$

Practically, at constant digging depth h = 200 mm, working efficiency is calculated as m^2/h, Fig. 3.39.

$$U = B\,v\,k_t / z \left[\mathrm{m}^2/\mathrm{h}\right] \tag{3.30}$$

B soil treatment width
v machine speed
k_t time utilization factor
z number of passes

As machine working speed depends on soil hardness, working efficiency will depend on machine working conditions in light, medium and hard soil category.

With B = 2.3 m, k_t = 0.8, z = 1, machine efficiency is as follows:

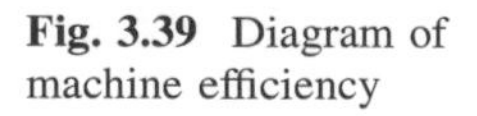

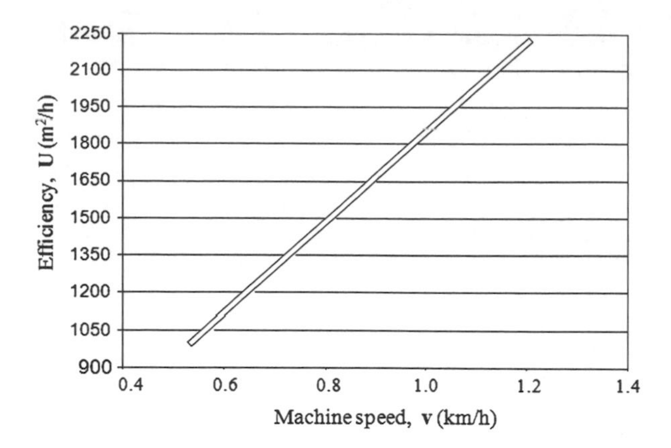

Fig. 3.39 Diagram of machine efficiency

3.4 Design of Heavy Demining Machine

Development of heavy demining machines aims to achieve great machine efficiency in demining operations in large and open areas. Using two working tools and wider soil digging should achieve higher digging velocity, so heavy demining machines typically use two or more tools for demining operations. The *Scanjack 3500* demining machine uses two flails. A combination of flail and tiller is used on *MV-20* demining machine, development of which is reviewed in this part, Fig. 3.40. Here, these issues are discussed in the case of a two flail, and for a flail and tiller machine [7].

The problem of neutralizing anti-personnel mines by activation or destruction with double tool involves determining the working speed of first and second tool. These parameters are adjusted so that mines of a particular size are destroyed by soil digging regardless of the relative position of a mine in relation to the hammer. For example, on a two-flail machine, the position of hammers and rotation speed of the first and second flail should be adjusted, to achieve the required density of soil digging. For the soil digging with two flails, working speed could be doubled and working efficiency increased. In practise, the machine user will determine technological working speed according to the estimated digging depth and mine threat.

3.4.1 Concept of Two Flails

Each flail digs a certain depth of overall soil layer, Fig. 3.41. It can be assumed here that each flail digs at 100 mm, which provides a total digging depth of 200 mm. Two rotor spirals, on which hammers are fixed, are joined symmetrically at the rotor centre.

Fig. 3.40 Heavy demining machine, MV-20 (Reproduced with permission from Ref. [10])

When determining the distance between flail shafts to avoid collisions between first and second flail hammers, chain elongation due to link wearing, is calculated, based on data MV machine [7], at 10 %.

Working speed

Density of hammer hits is determined by machine movement speed v and flail rotation speed n, as follows;

$$S = v/n \tag{3.31}$$

S	tool cutting feed (shear)
v	machine working speed
n	number of flail revolutions

Both flails have similar hammer positions and rotation speed.

φ_1	hammer angle of attack/grip of first flail
φ_2	hammer angle of attack of second flail
r_1	radius at which hammers of the first flail are hitting the soil
r_2	radius at which hammers of the second flail are hitting the soil
h_1	digging depth of first flail
h_2	digging depth of second flail

The highest soil clearance density is achieved at a hammer cutting feed and tiller cutting feed of 15 mm, but then the working efficiency of the machine is the lowest. Highest machine efficiency is achieved at higher machine working speed and at a hammer cutting feed of 30 mm. When operating a two-flail machine, power required to operate the first flail is usually different from that required for the second flail. The first flail digs the upper soil layer that can be covered with grass or other vegetation and can contain moisture, or in a winter, can be frozen. Thus the digging resistance of the first layer can be higher or lower than that of the

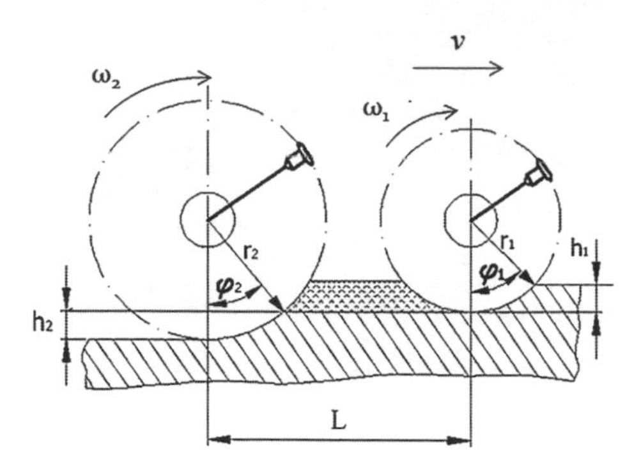

Fig. 3.41 Soil digging with two flails

resistance of second layer. Designers and operators must consider different combinations of digging resistance in the first and second layer.

3.4.1.1 Soil Digging Resistance

Soil digging is based on force impulse of the flail hammer. Depending on soil moisture, hammering cuts the soil and throws it behind the rotor. Existing here is the following assumption: if soil is dry and hard, then it is crushed and dispersed. Digging can be done on coherent soil, well bound and soft. Crushing can be done on the soil which is non-coherent, dry and hard.

Hammer digging resistance of coherent soil:

$$R_{ki} = z_n \, k_1 b \, \Sigma \, S_{ti} \, [\mathrm{N}] \tag{3.32}$$

z_n number of hammers in same digging position on the two hammer spirals
k_1 unit resistance of soil digging, depending on soil category
b hammer cutting width
S hammer feed on soil (16–30 mm) providing soil digging density for safe mine destruction
$\Sigma \, S_{ti}$ total thickness of soil digging layers ($S \, sin\varphi$)
φ hammer striking angle ($cos\varphi = r{-}h/r$)
φ_R angle between two hammers is a part of an angle which describes spirals on the over all rotor length.

Resistance to crushing of non-coherent soil by hammer:

$$R_{\sigma i} = \sigma \, A \, [\mathrm{N}] \tag{3.33}$$

σ soil strength limit
$A = z_n \, b \, \Sigma \, S_{ti}$ area of the hammer impact blade

3.4.1.2 Hammer Force Impulse

Impulse force during time interval Δt:

$$F_i = m_h v_o (1 - k) / \Delta t \tag{3.34}$$

m_h hammer mass on flail
v_o circumferential hammer velocity
k restitution coefficient
Δt force impulse time interval of hammer grasping of soil

$$F_{in} = F_{in} z_n \tag{3.35}$$

In order to realize the process of digging, the condition that the force impulse of the hammer is greater than the digging resistance has to be fulfilled.

$$\text{Soil digging condition}: \; F_{in} > R_{ki} \tag{3.36}$$

$$\text{Soil crushing condition}: \; F_{in} > R_{\sigma i} \tag{3.37}$$

3.4.1.3 Soil Digging Moment

At the start of operation both flails should be accelerated to working speed. During acceleration hammers are not in contact with the soil and power is used only to overcome inertial forces of rotor and hammers. Once the hammers achieve working speed, soil digging begins. It can be assumed that the soil digging moment, *Mk* is equal to the flail rotation resistance moment plus friction resistance because of hammer soil dragging, which causes hammer lag, as follows:

$$M_k = z_n J \varepsilon + M_d \tag{3.38}$$

M_k digging moment
z_n number of hammers in same digging position
J total inertial moment of hammers and chains
ε angular hammer velocity
M_d soil dragging moment

For this equation, the following assumptions have been made:

- soil digging is made only by hammers in the digging position (z_n),
- other hammers have constant circumferential velocity,
- cyclical operation of flails (deceleration or stopping with soil layer impact).

During design can be estimated the moment and power of hydromotor as follows:

$$M_{HM} = M_k / N\, i_p\, \eta_p \tag{3.39}$$

M_{HM} hydromotor moment
N number of hydromotors (two, at the end of the rotor)
i_p transmission ratio
η_p transmission efficiency

$$\begin{aligned} P_{HM} &= M_k \omega_{HM} \\ \omega_H M &= i_p \omega_M \end{aligned} \tag{3.40}$$

3.4.2 Flail and Tiller Concept

A working tool with independent positioning of a flail and a tiller is an innovative concept, which makes it possible to have different technological speeds of flail and tiller, as well as different digging depths. For destruction of mine threats the primary role is given to the first tool—the flail, Fig. 3.42.

Independent position and movement (in relation to digging depth) of flails and tiller provides higher efficiency than the classic two-flail arrangement with fixed relative flail positions. With this combination of flail and tiller certain advantages in digging of different types of soil can be achieved. According to demining requirements digging depth and number of rpm are adjusted in order to achieve higher technological speed for better working effectiveness. Each segment of tiller rotor can have multiple heads positioned at relative angle of 120°. Segments are phase shifted in such a way that segment cutting heads form three spirals, that start at the rotor centre and spread symmetrically to each side.

To neutralize AT mines, the primary role is given to the flail with its high radius; a secondary role is given to the tiller, which ensures digging depth and soil mine clearance quality (soil digging profile). The radius of the tiller is smaller than the flail so as to limit machine mass and ensure sufficient engine power, Fig. 3.43. Removal of mine threats of great destructive power has to be performed with the primary tool—the high radius flail—that will not be damaged. The tiller unit, as secondary tool, is adequate for demining depth adjustment and full mine clearance. Diagram of soil digging density with flail and tiller is given in Fig. 3.44.

To summarize, combining a flails as primary and tiller as secondary tool affords the following benefits (Fig. 3.45):

- different working conditions can be more flexibly handled by two independent working tools; possibility to independently adjust digging depth and rpm of tiller and flail; adjustment to real demining conditions
- two tools/tiller and flail can provide double efficiency

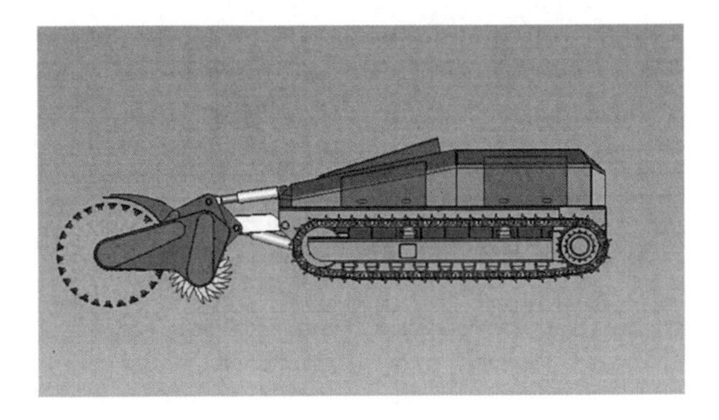

Fig. 3.42 Modelling of a demining machine, flail + tiller

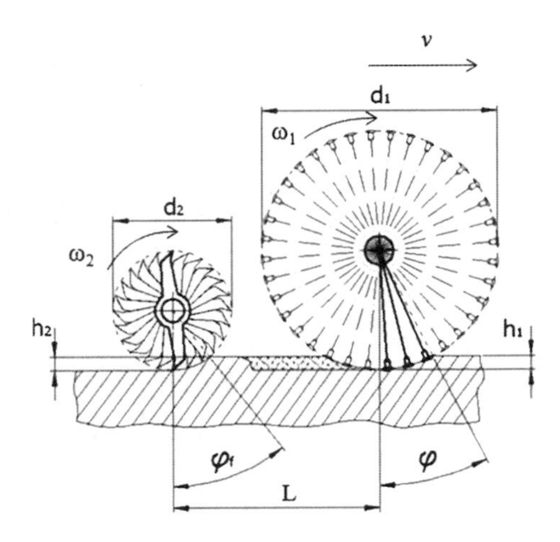

Fig. 3.43 Two independent tools—flail and tiller

- two different types of demining tools provides high reliability of mine destruction
- the most destructive AT mines are destroyed by flail, avoiding significant damage to the machine)
- the tiller is lighter than a second flail, affording materiel and engine power savings
- the tiller destroys the smallest parts
- the tiller can be adjusted to determine the final soil digging depth.

3.4.3 Machine Resistance on Mines

Overall machine protection includes protection of machine's sides and ballistic protection from mine fragments. Countermine protection is considered at two positions under the machine: the tracks and the area under the centre of the

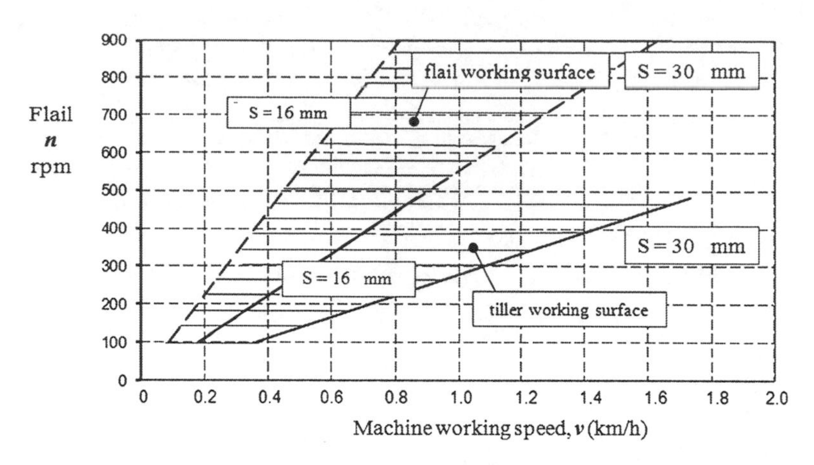

Fig. 3.44 Diagram of soil digging density, machine speed and tiller's and flail's rpm

Fig. 3.45 Operation with two independent working tools, flail and tiller, MV-20/MV-10 (Reproduced with permission from Ref. [10])

machine body. Overall protection from different threats is limited by machine's mass. Protection by square metre is being considered: *mine protection (kg/m^2) and ballistic protection (kg/m^2).*

Mine protected machines have to pass testing for AT mine explosions, which is necessary to measure the parameters of vehicle load impulses from basic AT mines with 6 kg of TNT for 2a and 2b protection level [2, 8]. In addition, it is very important to be familiar with load spectre and other mine threats, for example AP bouncing-fragmentation PROM-1 mine, including its potential effects on the operator that remotely control the machine from a distance of 100 m.

Test results of MV-20 demining machine according to HRN [2] (CROMAC CTDT—Test Site—*Cerovac, for details see Chapter VI*) are:

(a) Anti-personnel mines

AP mine types: PMA-1A, PMA-2, PMA-3, PMR-2A, PROM-1

At 5, 10, 15 and 20 cm digging depth, 55 % of mines were activated, 45 % crushed.

All mines were destroyed, no damage on the machine.

(b) Anti-tank mines

AT mine types and results: TMM-1/activated, TMA-3/activated, TMA-4/activated, TMRP-6/crushed.

TMA-3 mine is placed in front of the machine working tool at 5 m distance, depth of 10 cm and armed with fuses; Fig. 3.46. Machine and tools activate (flail and tiller) the mine. On the tools and the machine, there was no damage.

TMRP-6 mine is placed in front of middle axis of the working tool at a distance of 5 m, depth of 10 cm and armed with fuse for activation by stepping on. Machine and tool movement crash the mine. On the tools and the machine, there was no damage.

MV-20 technical data

Machine mass..........................35 t
Machine length..........................8.0 m
Machine length, without tools.......6.0 m
Machine width... ....4.0 m
Machine height ....2.50 m
Engine power............ ....665 kW
Working tools.....................flail, tiller
Working tool drive ....hydrostatic
Tracks width ....600 mm
Working width.........................3.40 m
Working speed ....2 km/h
Working capacity .... 2–3 ha/6 h

Flail type............................AM-020
Flail diameter...........................2.00 m
No of "mushroom" hammers.........82 pcs
Tiller type............................AF-020
Tiller diggers position..................helix
No of tiller diggers.....................82 pcs
Soil clearance density ....15–30 mm
Machine speed: ....6 km/h
Soil digging depth ....up to 40 cm
Machine steering ....remote control
Slope......................................37^0
Side slope.................................25^0

3.4.4 Conclusion

In the heavy demining machine project, two versions of working tools were considered: one with two flails and another with independently controlled flail and tiller. The function of each working tool was analyzed with the goal to achieve higher efficiency and independency of each tool in different working and demining conditions. Not so long ago, the combination and use of different demining tools, such as flail and tiller, was unimaginable, due to high requirements for power balance that had not even started to be considered. Nevertheless, with a certain allocation of power, a realistic possibility of such a combined solution was established for heavy and medium machine categories. Independent adjustment of two working tools and remote control speeds up machine demining. Machine testing according to HRN [2] that includes *Performance*, *Survivability* and

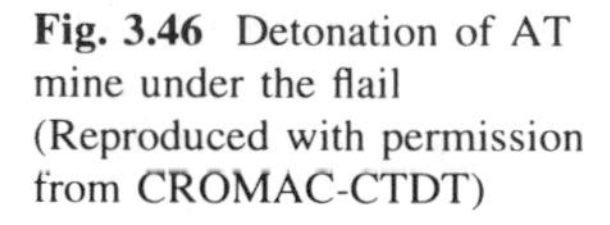
Fig. 3.46 Detonation of AT mine under the flail (Reproduced with permission from CROMAC-CTDT)

Acceptance, demonstrated high demining reliability and high quality of soil mine-clearing from all mine types, while providing insensitivity to damages from AT mine explosions.

To sum up:

1. Heavy demining machine has to provide high quality mine clearance and high working capacity in machine demining.
2. The most important parameter is the soil clearance reliability with two indepedent tools: flail, tiller. The tiller always guarantees a soil clearance depth, while the flail bears less damage when it activates AT mines.
3. During the machine demining process, machine should be remotely controlled from an escort vehicle. This provides operator safety and enables him to achieve high level of machine efficiency.

3.5 Support Demining Machines

Demining includes complex projects such as riverbanks, channels, forest roads, marshy terrains, etc. Hydraulic excavators have more and more important role in demining. Hydraulic excavators with long arm and special working tool are cleaning suspicious minefields on hard to reach terrains where other machines cannot be used. Excavators are used on riverbanks, dikes, channels, soft soil, woods, intersected terrain, etc. For that, excavators are using one of two available special tools: mine clearing flail or cutter for mid-size vegetation.

Development of special machines based on *Spider* type excavator, can be used in humanitarian demining of inaccessible terrains. They can be used in peace keeping operations for detection and clearing of explosive devices. Machines will be equipped with flail type demining tool, cutter for low and medium vegetation, and robot and other tools.

3.5.1 Demining Excavators

Based on previous experience in complex demining projects, most often used are excavators combined with light demining machines. To clear the approach part and prepare working area for excavators, light demining machines are used. After that, excavator with cutter is used to clear the vegetation, preparing the area for light demining machine that will dig the soil and destroy buried mines. If terrain does not allow use of the light machines, excavator will do the mine clearing, using flail that digs the soil and destroys buried mines. Operating in those conditions, excavators proved their high quality in clearing of different terrain according to following criteria: working efficiency, safety and work precision, terrain mobility, human protection and logistics. The advantages of demining excavator and flail are that they reduce the risk for personnel, reduce number of people required and are providing fast mine clearing method and control.

Standard tracked or wheeled excavators of 15–30 t are being modernized in accordance with user's needs in actual terrain conditions. Excavator arm can be extended for the required mine clearing reach (long-reach), thus increasing excavator working area. New designs for working tool, flail and vegetation cutters are developed. Hydraulic lines are placed along the arm and are protected against shrapnel. Additionally, by the use of armour excavator chassis is protected and operators cabin is replaced by armoured one. Driver-operator's safety is increased, being protected against shrapnel and AT mine explosion blast and noise, as well as rollover protective structure (ROPS) and falling object protective structures (FOPS). With reduced noise levels and use of ventilation, operator's comfort is maintained.

Demining technology is being constantly improved. Therefore, demining machine producers are modernizing their products according to new demining technology requirements. Modern technology of demining excavator design was proven in testing and real mine clearing environment. In order to do safer and faster demining, demining companies are more and more using modern excavators. From the effectiveness point of view, under difficult conditions, the best results have special excavators. Working capacity of such excavators is 2500–3000 m^2/day/5 h. Considering competition on demining market, it is hard to imagine that demining company does not use such excavators (one or more) either through procurement or lease. Either way, it is obvious that company competitiveness depends on excavator efficiency. In the most difficult environment, the price of demining per square meter is the highest. More and more companies are procuring special excavators.

Fig. 3.47 Hydraulic excavator with flail—long range, Liebherr *(Reproduced with permission from CROMAC-CTDT)*

3.5.2 Excavator Long Reach

Tracked excavator has extendable arm of 12.5 m (long-reach) or more if needed. Excavator mass is 25 t and has diesel engine of 135 HP. Excavator working mechanism provides for long reach, large working area, depth and height, Fig. 3.47. Mechanism can be adjusted according to specifications with mono-block arm or adjustable two-piece arm. Working tools: vegetation cutter or flail is quickly interchangeable. Working tools weight is 400–600 kg. Excavator working efficiency in mine clearing is 3000 m^2/5 h. Depending on working conditions, excavators have number of operating modes (LIFT, FINE, ECO, POWER/ lifting—slow speed, precision work, fuel saving at low power, digging in hard working conditions—maximum power). Excavator hydraulic system allows maximum power on flail or vegetation cutter (Fig. 3.48). For flail operation or vegetation cutter, additional engine of 125 HP is mounted.

Excavator efficiency:

- quick adaptation to complex conditions and large working area (long arm reach)
- safety and security of excavator operator
- high terrain mobility
- versatility of use

3.5.3 Operator Safety

3.5.3.1 Protection of the Driver-Operator and Machine Against Fragments of Bouncing-Blasting Mines

In controlled conditions, bouncing mine is laid laterally from the edge of the operating device at the machine height of about 70 cm. After mine activation, damage from the fragments can be noticed on the operating device, cabin and

Fig. 3.48 Demining excavator with vegetation cutter (Reproduced with permission from CROMAC-CTDT)

trailer amour. There must not be any damage trails inside cockpit (operator space). The safety cockpit windshield have to be impenetrable against the fragments, thus protecting the driver. It is common multi-layer glass, which can be quickly replaced in case of damage. On the inner side, the glass has to be ventilated to prevent from fogging during operation.

3.5.3.2 Protection of the Operator Against Impulse Noise and Shock Vibration

The power of the explosion impact noise (i.e. noise in dB) is determined on the bases of sound pressure (overpressure). Duration of the explosion sound pressure of an anti-tank mine TMA-3 is about 14 ms, and the peak level of the impulse noise amounts to around 150 dB (classical ear protectors, the so-called ear cups reduce the noise level by about 25 dB). Humans should not exceed the daily allowed noise values of the hearing organs (some dozen explosions). Shock vibration of the driver seat has to be below 15 g. Protected cockpit reduces the overpressure, and in combination with a helmet and a suit the overpressure on the ears is reduced to a tolerable limit. In order to achieve maximum noise reduction, interior of the armoured cabin is lined by sound-absorbing material. During development, studies were also carried out regarding the influence of impact force, i.e. acceleration/deceleration of the crew heads according to the allowed regulations. The amour protection of the mine clearing device is made of fragmentation plates of a certain thickness and angle.

3.6 Ecological Demining Machine

Beside industrial facilities, cars, trucks, agricultural and engineering machines are environment pollution actors, because they are using fossil fuels and lubricants for its operations and are producing exhaust gases such as carbon-monoxide (CO), carbon-hydrogen (HC), nitrogen-oxide (NO_x) and soot. According to EU directives, alternative fuels and lubricants are being developed. Additionally, high level of noise and vibrations is affecting human lives and work. Development and procurement of ecologically acceptable demining machines is proposed. Based on requirements and standards, model of ecologically acceptable demining machine is set up.

Perspective is given to alternative fuels and to new drive and transport technologies. *Characteristics of ECO generation* of demining machines are:

- environmental protection has become an important factor in the decision making of work processes and selection of means of work,
- demands requiring humanization of humanitarian demining work,
- many ecosystems are protected to a various degree.

ECO requirements

(1) *Bio-diesel fuel* to be used as engine fuel, and engine or hydraulic oil to be biodegradable (non-fossil) both for lubrication and hydraulic transmission systems, all pursuant to EU directives.

(2) *hybrid-electric propulsion*. During the last 10 years, electric transmission was developed, based on modern generator characteristics and electromotors with permanent magnets and intelligent power management system. Such vehicles provide advantages when mobility, usefulness, ergonomy, life cycle costs and ECO acceptability are concerned. In relation to classical drive, concept of energy tank and lower power engine, saves the energy and provides better machine efficiency. These projects are necessity for each country in order to gain necessary experience when design of vehicles with hybrid-electric drive is concerned. It can be expected that hybrid-electric vehicles and machines in next decade will be introduced into operation in developed countries. Introduction of these projects in the area of humanitarian demining technique leads to recognition and cooperativeness of each country.
It is important to open development projects in this area, in order to ensure technological basis for future production. Interesting systems for production are electromotor drive, energy converters with digital regulation system, etc. It means to master machine mechatronics and technologies for production of certain machine chassis assembly. This would create conditions for small and medium size companies to form production clusters in the area of humanitarian demining technique.

(3) *humanization of the workplace* in the cockpit by providing work conditions as close as possible to those achieved by several times heavier machines (more comfortable seats, countermine seats, double steering—joystick and steering-wheel-track, further improvement of the work environment and ergonomic characteristics, especially regarding noise, vibrations, climate conditions, etc.).

This generation of machines is a result of close relationship between society (state), science, (university, institutes) and economy, which is well known in current universal activities (s.c. *triple helix*). Preconditions for the success of these projects are adaptation of all participants and presentation of the new work to the market. Even though there is a great scientific potential that can be revealed on the market, one main challenge remains for the science and production systems: to direct large amount of scientific research knowledge to various economically sustainable products and services.

Machinery for demining operations cause a number of design problems which have to be solved in the course of their adaptation, e.g. selecting suitable engine, tires, tracks, transmission components, etc., strengthening the chassis and axles, securing a particular degree of stability, realizing the most favourable arrangement of graded gear box speeds, etc. Ergonomic indicator specifying humanization of the work is important. The only obvious problem is that it is located in a rather small cockpit space. The seat enables movements from left to right for half a circle, and different adjustments of the seatback. The most important purpose of measuring and evaluating vibrations is determining the effect it has on the operator's health during the machine operation. The next task is taking part in organization of equipment manufacturing, optimization of work regime, level of control, energy consumption, etc., as well as legislation of products (certification, occupational safety, typing…). It is always desire to combine power, economy, and environmental preservation in logging. For more efficient future generation machines, it is desirable to produce the equipment that will not damage the soil (roots of tree), which is saving energy and does not pollute the environment.

Machine engine, transmission gearbox, hydraulic components are all lubricated using specific biodegradable oil. Undoubtedly, it is necessary to replace fossil fuels that are difficult to degrade in protected areas with degradable bio-oils that are environmentally friendly to the forest. The machine engine can use fossil fuel like crude oil but also other renewable fuels colloquially know as bio-diesel fuels (produced from rapeseed, sunflower and similar oil sources as well as fuel derived from disposed food oil). Among other things, this can be considered as adapting the engine to renewable biodegradable fuel based on the EU strategy on reorganization of the EU energy sector. It began with publishing of the co-called *White Book* and adoption of a number of EU directives on the growth of share of renewable energy-generating products, for example, Directive 2003/30/EC on the promotion of bio-fuel in transport. Another reason for the enhancement and improvement of the machine was the significance of environmental protection and energetic efficiency.

(a) Diesel engine exhausts gases

Demining machines belong to the group of earthmoving machines. The equipment covered by the standard includes earthmoving machines, such as construction wheel loaders, bulldozers, highway excavators, and other like forklift trucks, road maintenance equipment, etc.

The first European legislation to regulate emissions from non-road (off-road) mobile equipment was promulgated 1997 (Directive 97/68/EC). The regulations for non-road diesels were introduced in two stages: *Stage I* implemented in 1999 and *Stage II* implemented from 2001 to 2004, depending on the engine power output.

Stage III/IV emission standards for non-road engines were adopted by the European Parliament, 2004 (Directive 2004/26/EC), and for agricultural and forestry tractors 2005 (Directive 2005/13/EC). *Stage III* standards are phased-in from 2006 to 2013, *Stage IV* enter into force in 2014. *Stage I/II* limits were in part harmonized with US regulations. *Stage III/IV* limits are harmonized with the US Tier 3/4 standards.

Stage III/IV emission standards

Stage III standards—which are further divided into two sub-stages: *Stage III A* and *Stage III B*—and *Stage IV* standards for non-road diesel engines. These limit values apply to all nonroad diesel engines of indicated power range for use in applications other than propulsion of locomotives, railcars and inland waterway vessels.

Stage III B standards introduce PM limit of 0.025 g/kWh, representing about 90 % emission reduction relative to *Stage II*. To meet this limit value, it is anticipated that engines will have to be equipped with particulate filters. *Stage IV* also introduces a very stringent NO_x limit of 0.4 g/kWh, which is expected to require NO_x after treatment.

Stage III A standards for non-road engines

Cat.	Net Power kW	Date[a]	CO g/kWh	NOx + HC	PM
H	$130 \le P \le 560$	2006.01	3.5	4.0	0.2
I	$75 \le P < 130$	2007.01	5.0	4.0	0.3
J	$37 \le P < 75$	2008.01	5.0	4.7	0.4
K	$19 \le P < 37$	2007.01	5.5	7.5	0.6

[a] dates for constant speed engines are: 2011.01 for categories H, I and K;2012.01 for category J

Stage III B standards for non-road engines

Cat.	Net Power kW	Date[a]	CO g/kWh	NOx + HC	PM
L	$130 \le P \le 560$	2011.01	3.5	0.19	2.0
M	$75 \le P < 130$	2012.01	5.0	0.19	3.3
N	$56 \le P < 75$	2012.01	5.0	0.19	3.3
P	$37 \le P < 56$	2013.01	5.0	4.7[a]	0.025

[a] NOx + HC

Stage IV standards for non-road engines

Cat.	Net Power kW	Date	CO g/kWh	HC	NOx	PM
Q	$130 \leq P \leq 560$	2014.01	3.5	0.19	0.4	0.025
R	$56 \leq P < 130$	2014.10	5.0	0.19	0.4	0.025

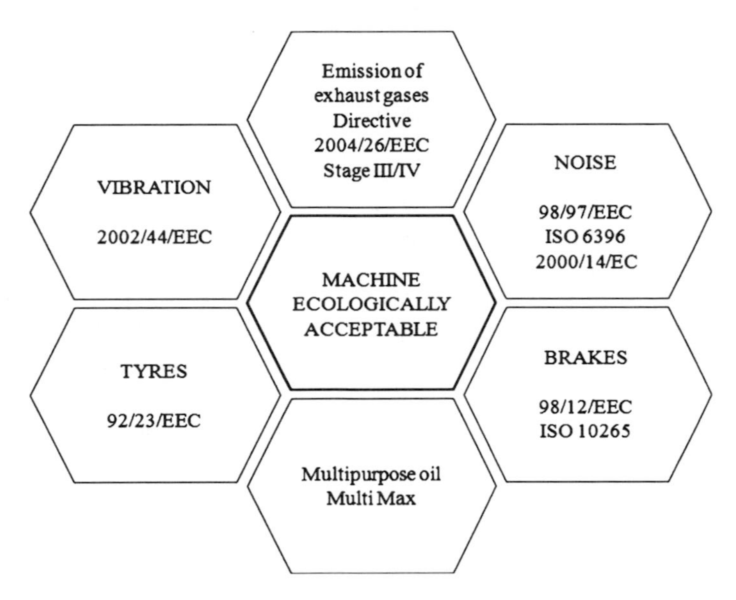

Fig. 3.49 An ecologically acceptable demining machine

3.6.1 Machine Acceptability Eco Model

Modern machines, used for soil processing, should fulfil requirements of *Stage III/IV* standards, allowed values of emission of exhaust gases for Diesel engines, and additionally, requirement guidelines for tyres 92/23/EEC, brakes 98/12/EEC, noise 98/97/EEC and vibrations 2002/44/EC. Noise boundary values for machines with engine power of 75 kW to 150 kW, is 72 dB to 80 dB, outside machine at distance of 7 m (92/97/EEC). In addition, it is important to respect allowed vibration level, to which machine crew may be exposed.

Selection of Diesel engine that provides lower emission of exhaust gases depends on injection and lubrication system. Modern system of direct fuel injection provides exact fuel quantity, under the exact pressure and crankshaft angle values, injects and distributes fuel into combustion space of diesel engine (*Common rail*).

Vibrations significantly affect human lives and work, human bones, muscles and complete neurological system. Directive 2002/44/EC, i.e. 89/391/EEC describe permitted vibration level exposure for machine operator. These are vibrations to which arms and body are exposed. Daily vibration exposure limits for arms are 5 m/s^2, and for human body is 1.15 m/s^2.

Hydraulic mineral oils have long list of advantages, but also a shortcoming—they are not degradable, and are polluting the soil and some countries have prohibited the use of mineral oils for agricultural and engineering machines. Multipurpose oils are solution for machine lubrication. Modern concepts are based on several oil types (3–4), which are used for engine lubrication, gearboxes, differentials, breaks and hydraulic systems. Future concept of MULTI oils provides universal oil, which fulfil lubrication and biodegradation standards. These oils are, through certain amount of time, degradable up to 99 %. Ecologically acceptable demining machine should fulfil, beside exhaust gases permitted level, following requirements: oil biodegradability, noise, vibrations, eco-tires and breaks, Fig. 3.49.

Alternative to ecologically acceptable machine is ecologically acceptable machine with diesel-electric drive.

(b) Diesel—electric and electric demining machine

Compared to hydrostatic concept, diesel-electric propulsion concept is regarded as technically more advanced. Development of the machine is in early stage of prototype demonstration and machine testing would provide basis for analytical judgment for acceptability of technical concept. Diesel engine is connected to the generator, which produces electricity transferred to the electromotor with permanent magnets, which is providing power for machine movement and soil digging flail. Initial calculations are implying towards better performance and higher power efficiency. Technological challenges of power conversion and high temperatures are expected to be solved in near future. Demining machine traction characteristics are sufficient for effective soil digging and for destroying AP mines. It is considered that more advanced technologies will enable multi role use of machine for humanitarian demining tasks and for special tasks of removal the remains of low intensity conflict. Electrical propulsion technology is intensively developed and distributed as a future crucial factor for new capabilities. The biggest advance in demining is achieved in field of machine demining.

In accordance with development goal for modern demining machines, this part describes concept and characteristics of future ecologically acceptable *MV-4DE* machine with diesel-electrical drive,[1] Fig. 3.50 [9]. Comparison with hydrostatic drive concept is the best way to evaluate a new diesel-electric drive concept. In developed countries, a concept of diesel-electric (DE) drive is being developed for commercial and military vehicles. DE drive offers better vehicle performance, better operability, easy spare parts production and lower maintenance cost. Diesel engine has a directly coupled electric generator, used as electric power source. Electric energy is transmitted by power electronics inverter system to the electric motors with embedded permanent magnets. Electric motors are connected to the tracks sprocket wheel through the gearbox or to the flail drive for digging the soil. *MV-4DE* machine is technically advanced providing a multipurpose application for humanitarian demining tasks and for special requirements in removal of low intensity conflict remains (LIC, Low Intensity Conflict).

The machine is tracked, with flail working tool that digs the soil destroys hidden buried AP mines. Flail consists of rotor on which chain-flails with hammers for soil digging are attached. In accordance with demining quality, machine has to treat the soil at required technological speed. During the machine movement, flail has to dig the soil down to the certain depth and at required density for safe neutralization of smallest mines.

Project requirements for development of 5 t DE demining machines are:

- Safe cleaning of AP mines down to required depth;
- Intensive use in hardest working conditions, soil category I to IV;
- Ecologically compliant machine, use of mineral oils and lubricants excluded;
- Diesel engine prime mover, optimal working point with low exhaust gases emission;
- Operating in extreme temperature conditions;
- Demining capacity (500–1000 m^2/h);
- Remote control.

Machine chassis has to be fit for different upgrades. Except basic machine version, chassis allows the use of different superstructure modifications, as machine for removal of suspicious obstacles and objects, robot for UXO removal, light reconnaissance engineering vehicle, amphibious vehicles, etc. Based on such a multipurpose program, necessary logistics for life cycle, economic profit and technological progress, sustainability and propulsive development is being imposed. Development of training simulators with future machine versions is part of scenario of this program.

This project has achieved the development of DE power transmission for machine movement and demining tool operation, development of vector control of

[1] Through this technology project of Ministry of Science and Technology, new diesel-electric propulsion technologies are being developed, in order to produce environment friendly demining machines. Machine development is in early stage of prototype testing. After factory testing, an in application field test will provide analytical and expert evaluation of technical solutions. Development and construction of this Croatian machine, was supported by STIRP project under the supervision of MZT of Republic of Croatia.

Fig. 3.50 Diesel–electric demining machine, MV-4DE (Reproduced with permission from DOK-ING)

DE magnetic field orientation with high-energy permanent magnets with digital signal processing, and development and production of generators and motors with permanent magnets. This goal results with development requirements for humanitarian demining machines. Tendency for future development is based on ecologically acceptable power transmission, which provides better power efficiency and competitiveness.

Development of diesel-electric drive is based on advantages in comparison to classic machine drive (elimination of hydrostatic or mechanical transmission, differential, gear-box):

- fast moment adjustment on drive wheels
- optimal operation of diesel motor with ecological characteristics
- easier and safer conduction of drive power and breaking
- automation by using drive-by wire system and DSP technology
- static conversion of energy with minimal losses
- simple production process, increased machine reliability and functionality
- lower maintenance and life cycle costs
- modern design with permanent magnets provides smaller dimensions and overall weight of assemblies and vehicle
- design flexibility—better space distribution
- better processing autonomy
- better vehicle performance
- lower fuel consumption
- lower exhaust gases emission
- optional: silent mobility and processing

Electric drive produces a constant torque through the wide range of shaft rotation speeds—from zero to full speed rotation. Applied electric motor drive is organized based on motor current and speed control loop. Problems noticed at power conversion by static converters are primarily high environment temperature, in which electronic components have to operate. The biggest challenge, assembly components procurement costs—will be solved in the nearest future through larger volume production of assemblies and product modularity.

3.6.2 DE Drive System

Diesel engine is equipped with a three-phase alternating current generator (380 V), which is directly connected to the engine shaft, Fig. 3.51. Generator's rotor acts as diesel engine fly wheel. Machine movement is achieved by two independent electric motors for each track, and two electric motors are driving demining flail shaft. Lifting and lowering of flail's arm is achieved hydrostatically by two hydro-cylinders. Diesel motor operates in optimal working conditions at the constant speed where generator achieves nominal power, rotating speed, frequency, voltage and current.

The required machine power consists of the power for operation and the power for machine movement. The amount of the losses for transmission, engine cooling, air filtering, secondary devices drive, etc. should be added to this power. The required power for machine operation and movement has been calculated, and amounts to around 150 kW. Out of this, more than 90 % of power is used for soil digging and the rest for the machine moving. Power distribution with this machine is very much the same. Also it is assumed that for safe flail operation at demining process, at least up to 75 kW is required per meter of shaft length.

As an option, DE drive can charge on board batteries to store energy and deliver it for silent mobility, silent watch and vehicle operating in accordance with specific NATO requirements. Energy for electric drive system is provided from generator or batteries or from both sources simultaneously, enabling machine movement and operation, and providing additional power in case of sudden increase of resistance. It means that machine has to operate in regular *electric mode,* which results in negligible thermal and acoustic emissions. Energy stored in batteries allows the machine to spread its role to the areas such as Special Forces, reconnaissance missions, etc.

Components of DE drive

Selection of electric motor for machine movement

One track start-up torque:

$$M_g = P_v \,/\, 2\,\omega[\mathrm{Nm}] \tag{3.41}$$

P_v—power necessary for machine movement

Electric motor nominal torque for one track:

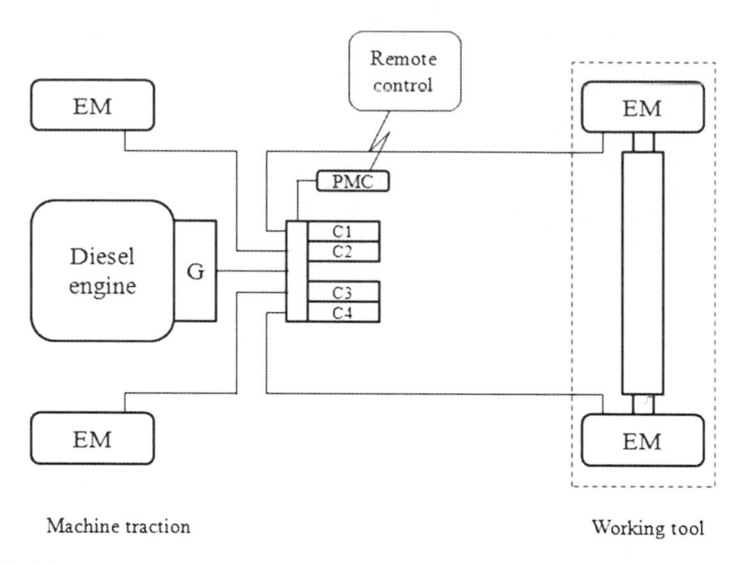

Fig. 3.51 DE drive basic scheme *G* generator, *EM* electric motor, C1,C2, C3,C4 controllers, *PMC* prime mover controls

$$M_{hm\,g} = M_g \,/\, i_{pr} h_{pr} \,[\text{Nm}] \tag{3.42}$$

Selection of electric motor for flail operation
P_r—power needed for soil digging
Electric motor moment for flail operation:

$$M_{hm\,m} = M_k \,/\, 2\, i_{lr} h_{lr} \,[\text{Nm}] \tag{3.43}$$

$M_k = P_r/2\omega$ [Nm]
i_{lr}—track sprocket wheel gearbox transmission ratio
η_{lr}—sprocket wheel gearbox efficiency

This is followed by the selection of the electric motor, nominal power, moment, and other characteristics.

Conclusion

The electric energy generated by the three-phase AC diesel generator is distributed to drive electric motors connected through the gearbox to track sprocket wheels and over chain transmission to the soil digging flail. Overall energy management and machine control is carried out by digital processing microcomputers connected over a CAN field bus in a common central control system.

Regarding development trends of hybrid machines, design of MV machine provides ecologically acceptable working machine regarding criteria for environment protection. Maintenance of DE power transmission is considered the

Fig. 3.52 Electric demining machine, MV-4ECO (Reproduced with permission from DOK-ING)

simplest, no need for oil change and dependency to oil quality. Traction characteristics of demining machine are sufficient for efficient soil digging and for demining and destruction of AP mines. Electrical drive technology is being developed for some time now and spreads as future factor for new capabilities. Technology of hybrid-electric vehicles provides advantages for special systems.

3.6.3 Electric Demining Machine

Developing problems are of technical significance and thus solvable. Capacity and service life of batteries, temperature conditions of use and need of transfer aggregate mounted on the supporting vehicle will represent logistical obstacles. With a technological solution of battery system with regard to stated 4 hour continuation work with incorporation of a supporting engine, could create certain utilization in humanitarian demining. The machine can be used to detect mine suspected areas and for demining of specific and smaller areas, Fig. 3.52.

References

1. International Mine Action Standards—Clearance requirements (2003), IMAS 09.10, 2nd Edition, United Nations Mine Action Service (UN MAS) New York.
2. Humanitarian demining—Requirements for machines and conformity assessment for machines (2009) Standard HRN 1142, Croatian Standards Institute, HZN 1/2010, Zagreb.
3. Test and evaluation of demining machines (2009) CEN Workshop Agreement, CWA 15044:2009, Supersedes CWA 15044:2004, CEN, Brussels.
4. Mikulic D, Koroman V, Ambrus D, Majetic V (2007) Concept of Light Autonomous Machines for Dual Use, Proceedings of the Joint North America, Asia-Pacific ISTVS Conference and Annual Meeting of Japanese Society for Terramechanics, University of Alaska Fairbanks, Alaska, Fairbanks.

5. Mikulic D, Koroman V, Majetic V (2006) Machine Demining Features. The 7th International symposium on technology and mine problem, Naval Postgraduate School Monterey, CA.
6. Mendek T (2004) Medium humanitarian demining machine calculation, Thesis, Faculty of Mechanical Engineering and Naval Architecture, Zagreb.
7. Mikulic D, Koroman V (2005) Development of Heavy Demining Machine, ISTVS, The international Society for Terrain Vehicle Systems, Proceedings of the 15th International Conference of the ISTVS, Hayama.
8. Protection Levels for occupants of Logistic and Light Armoured Vehicles (2004) STANAG 4569, Edition 1, Military Agency for Standardization (MAS), NATO/PfP, Brusseles.
9. DE drive system of MV-4DE (2002), STIRP, Brodarski institute, Zagreb.
10. MV-4 Mine Clearance System (2012) Catalogue, DOK-ING Ltd, Zagreb.
11. Mine Sweeper RMKA-02 (2012) Catalogue, Duro Dakovic, Special vehicles Inc, Slavonski Brod.
12. Mine clearing machine Samson 200/300 (2006) Catalogue, VILPO Ltd, Ljubljana.

Chapter 4
Design of Mine Protected Vehicles

Mine protected vehicle (MPV) is most common escort vehicle in demining operations. These vehicles are designed to protect the crew from mine threats in humanitarian demining with emphasis on cross-country mobility and countermine protection. Cross-country mobility implies high vehicle performances. Countermine protection implies full ballistic protection against bouncing—fragmentation AP mines and AT mines. Demining machine and other demining mechanization can be remotely controlled from this vehicle. Accordingly, MPV are very demanding project for designers and engineers.

There are specific mine resistant vehicles: *Casspir, Buffel, Tapir, Mine Killer, Mamba* (designed and developed by MECHEM, South Africa specifically for conditions in Africa), *Wer'wolf MkII* (Namibia). When steel wheels are mounted, these vehicles are used for AP mine destruction when treating suspicious areas. If chassis vehicle is seriously damaged, it is possible to replace damaged assemblies and parts in very short time.

In peacekeeping operations, ballistic and mine protected vehicles, called *All Protected Vehicles (APV)*, are used. These vehicles are, for example, *Dingo* and *Terrier*, from German manufacturer KMW—Krauss-Maffei Wegmann. *All Protected Vehicles* are designed to provide a safe and secure multi-purpose vehicle for peacekeeping and humanitarian operation. *Dingo* was based on a *Unimog* chassis produced by DaimlerChrysler, designed for high mobility on any terrain. Vehicle is of countermine design and protects crew against mine threats.

4.1 Project Requirements

MPV for humanitarian demining should be designed for multipurpose use with emphasis on cross-country mobility and countermine protection. Selection of standard 4×4 vehicle chassis of should be based on following requirements:

D. Mikulic, *Design of Demining Machines*,
DOI: 10.1007/978-1-4471-4504-2_4, © Springer-Verlag London 2013

- Total vehicle mass 10 t
- Maximum speed of 130 km/h
- Transporter/carrier (10 people)
- Version—citadel
- Cross country capability
- Countermine protection
- Ballistic protection
- Option to upgrade working devices
- Fast crew entry and exit
- Transportability: truck trailers
- Air transportability, C-130 H

Preliminary solution for vehicle design should be based on protection criteria according to STANAG 4569 [1]. Armour calculations for ballistic protection level II should be done accordingly. Additionally, *Armox* protective plates and safety glass should be selected, as well as isolation against sound and vibrations. Vehicles used for countermine tasks in humanitarian demining are specially designed vehicles based on highly mobile all terrain chassis. Crew protection and vehicle survivability should be evaluated too.

Crew and vehicle mass

- Maximum crew 10 crew members
- Vehicle mass up to 10 t

Basic characteristics

- Specific power 15–25 kW/t
- Maximum speed 130 km/h
- Slope and side slope 60 %/30 %
- Obstacles 0.50 m
- Fording without preparations 1.0 m

Chassis all terrain, 4 × 4 drive

- Engine Diesel engine, Euro 4/5, possible use of F-34
- Transmission automatic or semiautomatic gearbox
- Wheels safety tires 12.5 R 20, run-flat
- Brakes disk brakes, ABS and ASR
- Suspension independent wheel suspension

Armoured hull

- Crew safety compartment/monoblock, citadel-option
- Modular shape armour design
- Visibility, roof safety escape hatch, pressure valve
- Multi-layer safety glass
- 2–4 crew exits

Material

- *Armox* 400–500, 480–540 HB, thickness 4–6–8 mm
- /titan/hardox/kevlar/multi-layer glass

Mine threat level (MTL)/Mine level threats in humanitarian demining

MTL-01 (anti-personnel blast)
MTL-02 (anti-personnel fragmentation mine and small UXO)
MTL-03 A/B (blast anti-tank mines under wheel and hull)

Equivalent MTL 01/02/03, closest to the following protection standards:

(a) STANAG 4569

Countermine protection:

Level I AP explosive devices
Level II 6 kg TNT/
Level III 8 kg TNT

Ballistic protection 360^0

Level I (standard)/
Level II (option)

(b) EN 1063

- Protection class from B2 to B7
- Infantry weapons (rifles, pistols), protection in lower intensity conflicts

Crew's safety in cockpit/crew compartment

- Level II, explosion of AT mine, 6 kg TNT under the wheel,
- Impulse noise protection (max 140 dB),
- Acceleration protection (less than 10 g at anti-mine seat)
- Ballistic protection 360^0, Level I (standard)

Reparability, if damaged by AT mine activation
Users level maintenance (level I and level II)
It is considered that within the logistic concept the escort vehicle is to have a truck trailer to transport a light demining machine to the place of utilization.

4.2 Selection of Chassis

4.2.1 Cross-Country Mobility

Unimog chassis 4×4 is a very good choice for the design of Mine Protected Vehicles, Fig. 4.1. Important characteristics of the chassis include:

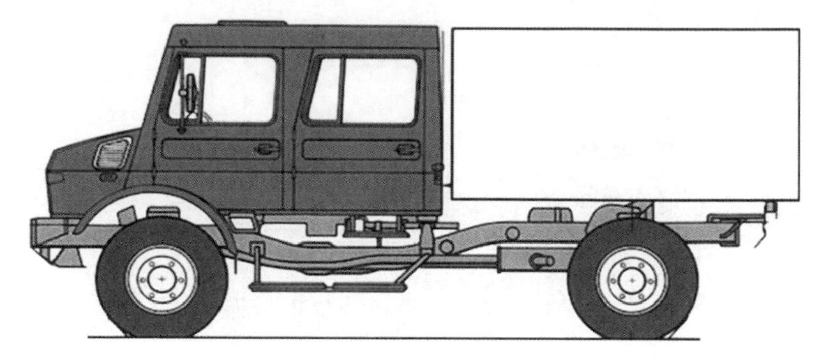

Fig. 4.1 Model of terrain vehicle 4 × 4, *MB Unimog* (*Source* Ref [4])

- highest clearance value of 500 mm and other cross-country mobility parameters
- highest vehicle load of 5 t, where highest load for the front axis is 4000 kg and for the rear axis is 4800 kg
- off road tow capability 2500 kg
- total vehicle weight may be up to 10 t with axle distance of 3250 mm
- U-profile chassis is acceptable for special vehicle superstructure

It is possible to equip the vehicle for different and multipurpose tasks. To transport 10 crew members, volume criteria is 1–1.2 m^3 per person, meaning that volume of 10–12 m^3 should be provided. Citadel option provides sufficient volume for 5–6 crew members.

It is possible to select engine power up to 200 kW with high torque at very low engine rpm (800 Nm/1200 min^1), providing high elasticity and usability in hard terrain and cargo conditions. Permanent four-wheel drive, with possibility for full differential blocking and change of specific pressure to the ground, provides high off road capabilities. If tire pressure decreases from 3.5 bars to 1.5 bars, tractive force increases around 40 %.

Basic chassis frame consist of two *U* profiles, joined by lateral tube profiles. Good characteristics of this design are payload, flexibility and resistance to torsion loads. Supporting frame can be mounted on basic frame, in order to prevent input and transmission of permanent deformations to the superstructure at extreme torsional stresses, while keeping all wheels on the ground. A vehicle of 10 t total weight can overcome slope resistance of up to 60 % on the asphalt, macadam and in mud. Vehicle maximum speed is 130 km/h for selected engine with power of 160 kW and torque of 810 Nm.

4.2.2 Selection of Chassis and Mobile Mechanic Workshop

Mobile mechanic workshops for maintenance of humanitarin machines are also mine protected vehicles. A mobile mechanical workshop can be planned also on a *Unimog* vehicle chassis of high soil trafficability, protected ballistically and counter-mine as escort vehicles. Due to transport height limitations of workshop vehicles the workshop ought to be projected as a variant of roof elevation. Therefore the workshop has two heights: transport height and working height. For a mechanic's work, workshop's internal height needs to be at least 2 meters. Workshop basic equipment comprises welding equipment, perforating equipment, turning equipment, cutting equipment, various tools, electroenergetic aggregates, and other.

4.2.3 Armour Material

Solutions for countermine and ballistic protection of special vehicles are including simple metal plates, spaced metal plates, ceramic/metal laminate and ceramic/composite laminate. In the past, standard solution was use of high resistance metal plates made of rolled homogenous armour (RHA, Rolled Homogeneous Armour), hardness of 350–380 HB. Most of wheeled vehicles have high hardness steel ballistic plates of (HHS, High Hardness Steel), of 480–540 HB.

For mine protected vehicles, there are highly hard steel plates (up to 600 HB), for example type *Armox*. On combat vehicles, such a plate is mounted in front of the main armour. Alternative solution is mounting the ceramic plates on the outer side of main armour plates, considering that ceramic is much harder than steel (2000–3000 VHN versus maximum of 750 VHN, measured according Vickers). More developed approach includes use of *composite materials,* usually consisting of external ceramic layer (Alumina, aluminium oxide), placed on multilayer supporting element made of aramid fibres (such as *Kevlar, Twaron*) or glass fibres (such as E glass), joined in matrix by thermoplastic resin. Kevlar is the registered trade name used by *Du Pont* for the aramid fibre. Kevlar is a flexible fibre, it can be woven into cloth, such cloth is strong enough to stop high-speed projectiles and—in layers-will stop bullets. Kevlar does not burn or melt. Ballistic protection of vehicles, Level I, II, or III (STANAG 4569) can be built by use of laminated composite armour, for example *FLEX-PRO Armour.* At NIJ standard 0108.01 Level III-A configuration of this material has little aerial density that provides a significant reduction in weight of vehicle. Composite *materials* have lower aerial density (60–65 % lower) than steel armour plates, but provide the same protection level. Impact into external ceramic layer, crushes or weakens the projectile, and resistance energy is efficiently absorbed and scattered through processes of dynamic deformation of supporting layer. However, composite armour plates cannot be used for load bearing structural elements, but only as additional armour.

Fig. 4.2 Countermine and ballistic protection design of an escort vehicle (*Source* Ref. [5])

4.3 Mine Protected Vehicle Design

4.3.1 Safety Compartment

The safety compartment/citadel is protected against mine explosion by multi-layered undercarriage structure. It consists of a sandwich construction, which is designed to absorb fragmenting mine blasts and can be reinforced to protect against anti-tank mines by fitting the blast shield. Key to surviving the mine blast is efficient absorption or dissipation of energy. Complete mine protection subsystem consists of the following elements [5]: deflector system, crew safety compartment with false floors, blast mounting for deflector and crew compartment, damping elements for the seats, 4-point safety belts. The mine deflector system is also specially developed to significantly increase mine protection of light, wheeled vehicles. The deflector consists of a structure of materials as well as a special blast mounting to the chassis. It serves to absorb a significant amount of the blast energy by mechanical deformation (Fig. 4.2).

Instead of the original driver's cab, MPV should have a citadel/safety compartment. Citadel can be especially designed and is produced in a special technology, called "Thin-plate-bending", eliminating most of weldings. The compartment should have a large, armoured front window providing an outstanding field of view for the driver, also equipped with armoured glass (1 or 2 parts). The resistant roof on the hardtop version provides protection against grenade

Table 4.1 Ballistic body armour levels

Level	KE—threat	Artillery
1	Assault riffles Ball ammunition Distance: 30 m Angle: azimuth 360°; elev. 0°-30° Ammunition: 7.62 mm × 51 NATO ball, velocity: 833 m/s 5.56 mm × 45 NATO ss 109, velocity: 900 m/s 5.56 mm × 45 NATO ss 193, velocity: 937 m/s	155 mm Estimated range of burst 100 m angle: azimuth 360° Elev. 0°–18°
2	Assault Riffles Armour piercing steel core Distance: 30 m Angle: azimuth 360°; elev. 0°–30° Ammunition: 7.62 mm × 39 API BZ; velocity: 833 m/s	155 mm Estimated range of burst 80 m angle: azimuth 360° Elev. 0°–22°
3	Assault and Sniper riffles AP tungsten carbide core Distance: 30 m Angle: azimuth 360°; elev. 0°–30° Ammunition: 7.62 mm × 51 AP (WC); velocity: 930 m/s 7.62 mm × 54R B32 API; velocity: 854 m/s	155 mm Estimated range of burst 60 m angle: azimuth 360° Elev. 0°–30°

(*Source* Ref. [1])

fragments. Most of the armour protection should be modular, adapted to the basic structure of the safety compartment.

4.3.2 Crew Protection Standards

Equivalent to mine threats in humanitarian demining (MTL 01/02/03) may be STANAG 4569 Anex A (Table 4.1), Anex B (Table 4.2) for logistic military vehicles, as well as an option of ballistic protection against mine fragments—EN 1063 standard. For mine protected vehicles used in humanitarian demining, level II for countermine protection (Annex B) may be used, on the basis of "cone of mine destruction" under the wheels and under the centre of the vehicle.

Mine threat levels in humanitarian demining

MTL-01, anti-personnel blast
MTL-02, anti-personnel fragmentation mine and small UXO
MTL-03 A/B, blast anti-tank mines under wheel and hull
MTL-04, medium-size UXO
MTL-05, anti tank HC Mines
MTL-06, anti-tank SFF Mines SFF—Self Forging Fragment
MTL-07, UXO Heavy

Table 4.2 Floor protection levels for logistic and light armoured vehicle occupants for grenade and blast mine threats

Level		Grenade and Blast Mine Threat	
1		Hand grenades, unexploded artillery fragmenting submunition, and other small anti personnel explosive devices detonated anywhere under the vehicle.	
2	2a	Mine Explosion pressure activated under any wheel or track location	6 kg (explosive mass) Blast AT Mine
	2b	Mine Explosion under centre	
3	3a	Mine Explosion pressure activated under any wheel or track location	8 kg (explosive mass) Blast AT Mine
	3b	Mine Explosion under centre	
4	4a	Mine Explosion pressure activated under any wheel or track location	10 kg (explosive mass) Blast AT Mine
	4b	Mine Explosion under centre	

(*Source* Ref. [1])

EN standard, protection class B2–B7, defines ballistic vehicle protection. Correlation between European (EN), US (NIJ), Russian (RSS) and German standard (DIN) is as follows:

NIJ	EN	DIN	RSS
I	B1	I	1
II–IIIA	B2, B3	I	2
–	SG1, SG2	II	2a
–	B4	II	3
III	B5	III	4
–	B6	–	5
IV	B7	IV	6

4.4 Level of Ballistic Protection

For design of level II protection for MPV, *Armox* steel plates of 4–6–8 mm thickness can be selected, which will be additionally coated to prevent fragments scattering towards the crew. Key parameter for MPV design is armour specific mass in relation to the protection level.

Figure 4.3 presents standard specific mass for four different materials for standard protection level [2]. It may be noticed that steel weight for protection level II will be about 80 kg/m^2, and for level III around 100 kg/m^2. Windshields, windows and other glass surfaces are usually made of ballistic glass, consisting of several glass layers, with protective membranes between layers, in order to prevent fragments penetration into the vehicle. For protection level II, required glass mass is around 115 kg/m^2, and for the level III is around 180 kg/m^2. New type of ballistic glass allows thickness to be decreased for 25 % for the same protection

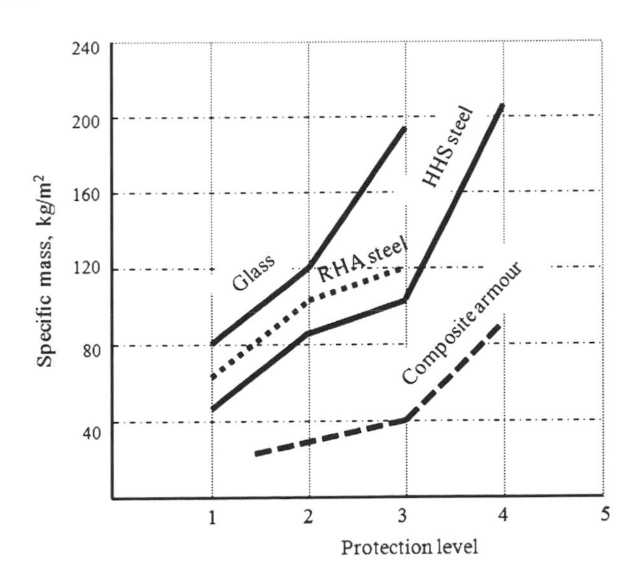

Fig. 4.3 Ballistic material mass in relation to protection level (*Source* Ref. [2]) *RHA* Rolled Homogeneous Armour, *HHS* High Hardness Steel, Composite armour (alumina + fibre laminate)

level. Armour mass evaluation for transporter type based on terrain 4 × 4 vehicles is shown in Table 4.3.

Characteristics of Armox 500 T plates

Armox 440 T and *Armox* 500 T type steel plates have good countermine and ballistic characteristics as materials of high hardness of 480–540 HB [3]. *Armox* plates can be mechanically worked and bended, Table 4.4.

Safety glass

Depending on protection level, highly resistant polycarbonates, acryl and glass are combined, bonded by multi-component special adhesives. The result is "glass", which is lighter than usual products of same protection level.

Glass protection principle

When bullet hits the acryl (fragile, but of high strength when reinforced and installed in layers), it absorbs bullet energy. Polycarbonate layer is elastic, and absorbs the rest of the bullet energy and prevents penetration into the vehicle. On the outside, damage in a shape of "spider net" can be found, but on the inside,

Table 4.3 Armour mass evaluation for transporter type based on terrain 4 × 4 vehicles

Countermine protection $\delta = 6$ mm	9 m^2
Ballistic protection $\delta = 6$ mm	12 m^2
Roof protection $\delta = 6$ mm	7 m^2
Total armour surface	28 m^2
Total protective glass surface	2 m^2
Total armour weight, min	1,550 kg
Total armour weight, max	2,470 kg
Vehicle chassis weight	4,000 kg
Total weight of armoured vehicle, min 2nd level	5,550 kg
Total weight of armoured vehicle, max 2nd level[a]	6,470 kg

[a] As maximum vehicle load is 10 t, it means that with 10 crewmembers included (1000 kg); payload of the vehicle will be 2–3 t, which can be considered as sufficient

Table 4.4 Characteristics of *Armox* 500 T plates

Chemical structure:	C_{max} 0.32 %; Si 0.1–0.4 %; Mn_{max} 1.2 %; P_{max} 0.015 %; S 0.01 %; Cr 1 %; Ni 1.8 %; Mo 0.7 %; B 0.005 %
Mechanical characteristics:	Hardness 480–540 HB Charpy test 20 J at −40°C tensile strength $Rp_{0.2}$ 1250 N/mm^2 Yield strength Rm 1450–1750 N/mm^2
Plate dimensions:	4 × 2,000 × 5,000 mm 6 × 2,400 × 6,000 mm

(*Source* Ref. [3])

glass is smooth and undamaged. Safety glass matches protection levels III and IV, according to NIJ standard. Glass can be used for doors and windows, combined with impenetrable armour.

4.5 Cone of Destruction

When designing vehicle floorboard, spectrum of loads affecting vehicle and exposed surfaces, caused by mine explosion under the wheels (level 2a test), under the vehicle centre (level 2b test) and on the sides of the vehicle, is shown in Fig. 4.4. This provides basis for MPV calculations, in order to provide crew protection against fragments, noise and blast vibrations.

Mine protected vehicles have to pass experimental verification done by the explosion of reference AP and AT mines. First, parameters for the change of vehicle load caused by AT mine explosion of 6 kg TNT or suitable mine dummy (for example, Nitropent charged) have to be determined. Additionally, it is important to be familiar with load spectrum for "Cone of destruction" and other mine threats on humans and on vehicle (AP bouncing—fragmentation PROM-1 mine).

Blast pressure under the vehicle

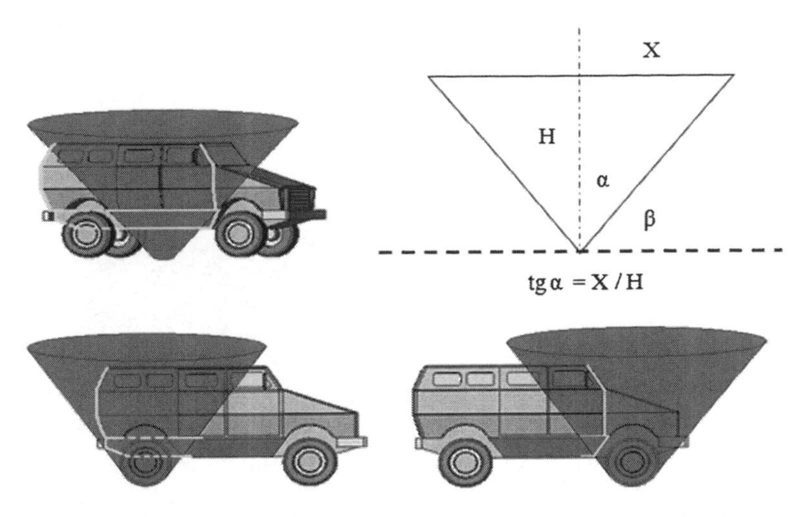

Fig. 4.4 *Cone of destruction* of the MPV, mine explosion pressure activated under centre (2b level) and under wheel (2a level)

Name/type of explosive, 6 kg TNT

Reflected pressure, «Side-on» pressure, detonation products pressure

p_{ur} reflected blast pressure

p_{us} blast wave "side on" pressure

p_d detonation products pressure

X_i distance above detonation point

Based on diagram, it can be seen that for vehicle clearance of 50 cm, reflected blast pressure *(p_{ur})* of AT mine of 6 kg TNT is around 600 bars, Fig. 4.5 [6]. For protection against this blast wave, usually double floor and V-shape of floorboard sides is used.

If wheel runs on and activates AT mine of 6 kg TNT, detonation products pressure is responsible for wheel destruction. At distance of 1 cm, detonation products pressure is around 80000 bars, and if distance increases to 50 cm, effect of detonation products pressure is almost completely lost.

Accordingly, for floorboard design, reflected pressure *(p_{ur})* is dominant under the vehicle, and under the wheels dominant is detonation products pressure *(p_d)*. Blast wave side on pressure *(p_{us})* at distance of 50 cm is around 200 bars.

4.6 Verification of Countermine Protection

Approach

The goal is to test if vehicle fulfils countermine protection requirements, primarily for level 2a. To develop realistic simulation test conditions, during the design

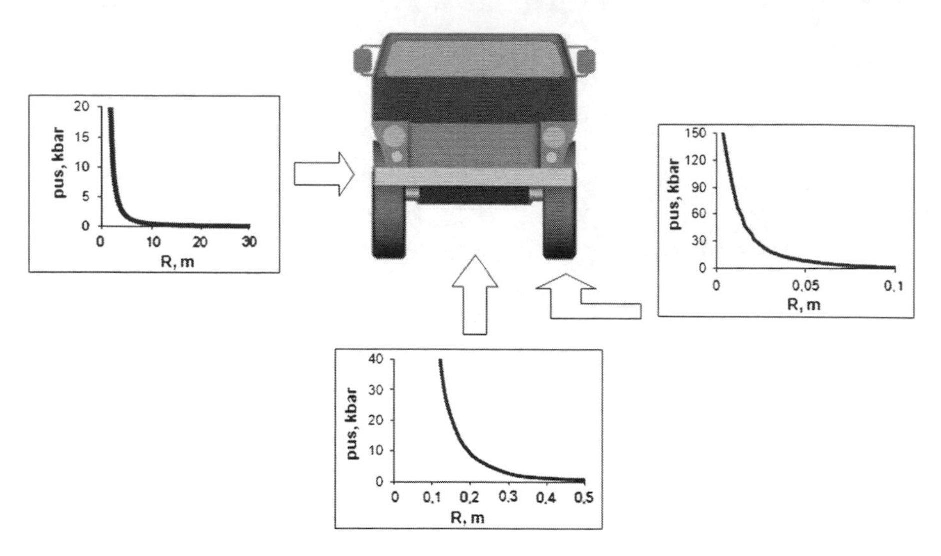

Fig. 4.5 Evaluation of MPV protection from shock wave pressure of AT blast mine explosion under the vehicle, wheels and at vehicle sides

phase, section module can be developed, consisting of front wheel and related armour parts for crew protection above the wheel. Wheel section module can be prepared and tested against mine explosion, or it can be inserted into body of old, used vehicle, Fig. 4.6. Wheel is lifted above the ground and chained to the vehicle body, in order to overcome suspension, shock absorber and spring resistance. This enables setting up the steel pits with mine charge under the wheels (or tracks in case of tracked vehicles).

Setting up the mine charge in accordance with evaluation of worst case scenario "cone of destruction"

Mine is placed under the wheel, offset to the inside, with 50 % of mine width, at most, is under the wheel, and the following relation apply [5]:

$S/2 \leq d \leq 40\ \%\ (S + D)$

S wheel print width

D distance between mine centre and tire

D mine charge diameter

For simulation purposes, instead of buried AT mine of 6 kg TNT, mine surrogate, based on NITROPENT charge (so called PENTHRITE—PETN) and set inside the steel pit, can be used. PETN density is 1.45 g/cm^3. AT mine of 6 kg TNT is equivalent to 5.1 kg PETN charge of certain dimensions. Explosive is initiated on the bottom of the charge, and initiation point can be located at most as 0.33 of explosive charge height. Ratio between height and diameter (H/D) can be of 1/3 with tolerance of up to +5 %.

Module configuration contains serial production axle and complete wheel suspension system connected to chassis and armoured vehicle body. To test thickness of armour protective section plates, plates are made of 6 mm/8 mm of thickness, for example *Armox* 440 T.

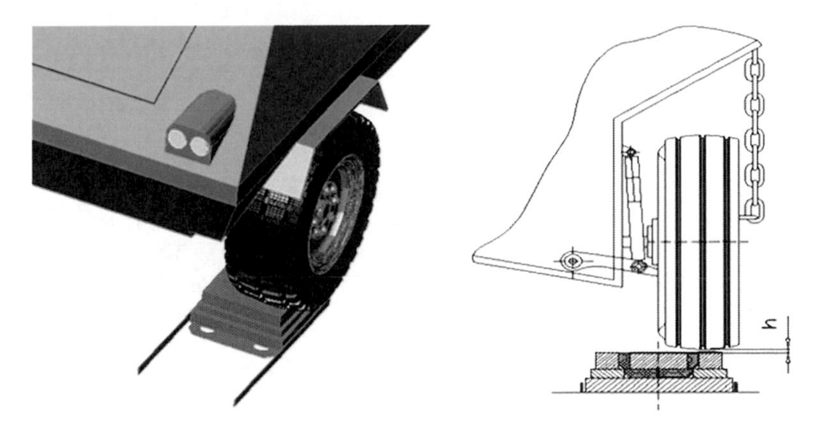

Fig. 4.6 Front wheel test configuration

Survivability of the MPV

After remote activation of mine charge, results of successful design of front protective vehicle's section can be seen, in accordance with survivability requirements. Damages and protective module condition are being checked, positions where axle parts hit the body armour, and armour plates and welds deformations.

Reparability of the MPV

Important MPV characteristics is reparability after being damaged by AT mine detonation under the wheels, which is in accordance with set requirements for quick vehicle's repair by replacing wheels and axles. Vehicle user should replace wheels and axles on the spot, which can lower the reparation costs. If mine detonates under the vehicle, regarding chassis V-shape, double floor and high clearance, replacement of pressure/explosion deflectors under the vehicle's armoured body.

Elements of the survivability test

- different levels of protection for various mine threats
- provides acceptable injury level for the crew—no injury at all
- provides crew safety in real conditions
- vehicle, components and assemblies durability
- reparability, repair costs

4.7 Protection Against Impulse Noise and Vibrations

4.7.1 Survivability of the Crew

Intensity of explosion blast wave (i.e. impulse noise expressed in u dB), is determined based on sound pressure (overpressure). Highest noise level should not exceed allowed values for hearing organs. For overpressure, higher than 200 Pa (140 dB), protective equipment have to be used (ear protectors, protective helmets). External vehicle parts influence noise level, i.e. thickness of armoured plate affects the reduction of noise level within cockpit. Duration of impulse noise close to the vehicle, i.e. one AT mine explosion, is about 14 ms, and peak noise value is about 160 dB. Thereby, ear protectors lower the noise level for 25 dB, which can be considered as well protected. Shock wave may damage unprotected ears, but also internal organs filled with air. See Table 6.1, for crew safety—no injury levels for the ear, foot/ankle and spine.

Vehicle design includes designing of external and internal protection. External armoured protection, made of armoured plates, primarily lowers the energy of fragments. Internal protection includes seats for isolation against vibrations and coatings for protection against noise. Internal cockpit isolation against noise includes also isolation using absorptive materials sound barriers.

Sound isolation using absorptive materials usually made from polyester or polyurethane foam of different thickness and density. Material ''catches'' the air into its structure and prevents its further expansion. When sound wave enters into structure, it disperses, and material absorbs energy of vibrating air molecules. It also prevents air rebounding from surfaces. This is most widely used and cost-effective solution that can be built on additionally.

Isolation using sound barrier achieves the same effect, but in a different way. Sound wave reaching the barrier is rebounding and turns back as an echo. Materials used for sound barriers are mostly rubber or mixture of rubber and polymers. They are denser and thinner than absorptive materials. Additionally, they are more expensive than absorptive materials, and are used when more durable material is required. Beside sound isolation, sound barriers may be used as isolation against vibrations, when used on metal sheets. Base material, found between isolation materials and the surface, is most often made from natural or synthetic rubber covered with sticky resin, oil and antioxidants. Beside its purpose as isolation material, it provides good adherence on surface that has to be isolated.

4.7.2 Protection Packages

Inadequately protected vehicle surfaces and vehicle devices, can be additionally ballistic protected, using certain protection packages (i.e. minimum III A level 32 layers g/m^2). Such ballistic packages can be made of *aramide fabric (Kevlar,*

Twaron CT-709). Using *aramide* protection, floorboard, seats, doors, equipment and other elements can be additionally protected. Ballistic packages are coated with different materials, e.g. textile, plastic, rubber or other, for waterproof protection and fixing on appropriate surface.

Blast wave that affects the vehicle's body, creates vibrations and sudden changes of operator position, which can further cause complex injuries of foot and ankle, if legs are leaned against the floor, as well as spine injuries due to vertical seat movement. High exposure to vibration blast increases the risk of spine injuries and pain in lumbal part. Extremely high blast values, such as movement on uneven and rugged terrains or explosion under the vehicle, can cause spine fractures. Additionally, long-time exposure to lower blast values causes degeneration of spinal disk and spine itself, resulting in constant pain. Due to high changes of vertical acceleration in short time, operator has to sit properly, because sitting properly lowers loads on spine as well as loads on overall musculature.

Seats that are built in into countermine vehicle should protect man from explosion blast vibrations (at least 10 g). In addition, seat has to provide stabile position and has to be designed according to ergonomic principles; seats have to reduce vibration as much as possible. Seat design has to endure all stretching that can emerge in its exploitation, and to withstand forces in case of eventual overturning. Seats have to be equipped with safety belts or adequate devices that will restrain unnecessary personnel movement and will firmly hold personnel back in their seats in case of vibration blast but will not limit necessary movement for steering the vehicle or operation of seat suspension. Today, we have specially developed "suspension seats" as alternative to classic seats, which are suspended on elastic mounts, avoiding direct transmission of blast to the crew. Seats are providing protection from injuries from AT mines explosions under or at the side of the vehicle and, as well as in case of overturning. Seats have incorporated molded padding and five point seat harnesses to reduce sudden body movements caused by both the primary (rising) and secondary (falling) effects of blast. Specially designed wrap-around headrests provide protection from whiplash. The MPV achieves this through a series of design features. Large tires fitted with run-flat inserts provide absorption; the fitting of deflectors along the wheel arches enhances blast pressure dissipation. Another key protection element is the high-performance collective overpressure NBC-protection system, which is identical to systems used in combat vehicles.

Active seat model

More complex seat model uses damping elements which are based on rheological features of fluids, MR dampers (Magneto-Rheological). Principle of MR fluids is based on oil containing micro particles that can be magnetized. In regular conditions (no load, no blast, no vibrations and no magnetic field) these particles are in free motion, in the oil that has density similar to engine oil. Exposure to magnetic field transforms fluid into solid within milliseconds. It can be easily turned back into liquid, when exposure to magnetic field ends (it changes its rheological features). Transformation level of MR fluids is proportional to the value of magnetic field to which it is exposed. This is nonlinear system, and optimization and calculation is more complex than in aforementioned linear systems.

Conclusions

A MPV type escort vehicle is to provide a clear cabin overview for remote control of the demining machine and safe protection for machine operators at a distance of 30 m. Vehicles have to be designed and tested according to safety standards for NATO logistics vehicle.The most important parameters of MPV vehicles are the distance between soil and the axle, double floor, deflectors, independent wheel suspension, run-flat tires, countermine seats, impulse noise and vibration protection.

References

1. Protection Levels for occupants of Logistic and Light Armoured Vehicles (2004) STANAG 4569, Edition 1, Military Agency for Standardization (MAS), NATO/PfP, Brusseles.
2. Bianchi F (2002) Protected Carriers for New Roles (II), Ballistic Protection of Light Tactical Vehicles, MILTECH 10, Monch Publishing Group, Bonn.
3. Technical specification Armox Armour Plate (2004) Ssab, Oxelösund AB.
4. Unimog Technical Manual U 4000 (2003), Stuttgart.
5. Mikulic D, Stojkovic V, Gasparic V (2005) Modeling of All protected Vehicles, 4th DAAAM International Conference on Advanced Technologies for Developing Countries, Slavonski Brod.
6. Suceska M (1999) Calculation of detonation energy from EXPLO5 computer code results, Propellants, Explosives, Pyrotechnics 24/1999.

Chapter 5
Personal Protective Equipment

In humanitarian demining one differentiates the following kinds of Personal Protective Equipment (PPE):

- personal protective equipment for deminers (manual demining), and
- personal operator equipment of the personnel involved in demining activities (machine demining, demining machine operator, driver of the escort vehicle, supervisors, support workers,…).

In the work of a deminer and the machine operator, their mobility proves highly important. A growing level of protection implicates greater weight of equipment, whereby reducing their mobility. Thus, a balance between the level of protection and the possible threat and working conditions is needed. In contrary, weaker capabilities of personnel arise, leading to lower efficiency and can add to greater accident probability.

Personal protective equipment of a deminer

In manual demining, deminers are located in a zone of greatest risk and must be eqiupped with quality PPE that will prevent head and body injuries in case of an AP mine explosion. Evaluation of personal protective equipment in humanitarian demining is conducted according to the *CEN Agreement CWA 15756, Personal protective equipment—Test and evaluation* [1]. Equipment refers to protection against AP blast mines, but does not apply to protection against AP fragmentation mines.

From the safety standpoint, design and characteristics of a deminer's protective equipment have to fulfil the following requirements:

- Protect vital body parts against AP mine explosion effects,
- Provide adequate body mobility and freedom of movement depending on working position (standing, kneeling, lying, light weight and easy to dress and undress),
- Use in hot and cold environment

D. Mikulic, *Design of Demining Machines*,
DOI: 10.1007/978-1-4471-4504-2_5,

PPE equipment is of modular type, and contains the following:

- Protective suit (vest, sleeves, trousers),
- Protective helmet with visor,
- Protective footwear (boots, boot-type slippers).

5.1 Personal Protective Equipment of the Machine Operator

One can distinguish the following kinds of personal operator equipment:

- PPE for the drivers of escort vehicles and machine operator controls the demining machine remotely from the escort vehicle,
- PPE of a machine operator moving behind the machine and who remotely controls the machine (light and medium demining machines),
- PPE of a machine operator who directly controls the machines from the machine cabin (usually, heavy demining machines).

The demining machine sometimes cannot be controlled from an escort vehicle, due to various obstacles. Then the machine operator moves behind the machine. During the machine demining of mine-suspected area, embedded mines are neutralized (destroyed or activated). However, occasionally some random AP mines or their parts may remain, presenting a threat of injuries. Therefore, the machine operator remotely controlling the machine, needs to be protected from an accidental explosion of AP blast mines. Of primary concern are legs, body (torso), neck and head. Operator legs, are at the moment of the explosion, exposed to accelerating forces that result from a combination of shock wave pressure and blasted mine fragments. According to the model "cone of destruction" of an AP mine, if an operator is too close to the explosion of an AP mine, there exists a great risk of injuries, ranging from feet and knees to hips and the genital area. Protective machine operator equipment can prevent huge injuries if legs are further away from centre of the "cone of destruction".

Except the above mentioned, protective equipment safeguards the operator from the thrown rock fragments and other debris which the working tool can blast of far away from the machine. Kinds of machine operator's personal protection:

- Passive operator protection
- Active operator protection

Passive protection refers to procedures of the machine operator during the very process of demining. When control from an escort vehicle doesn't seem possible (smaller spaces for demining, cross-country, forest, and other), the operator moves outside the vehicle. In this case the operator takes on passive protection:

- keeps a safe distance from the machine in relation to the possible threat (50–100 m considering the estimate of the AP fragmentation mines),
- the operator walks behind the machine, following the rut of the machine's tracks or wheels, whereupon the machine silhouette provides a shield for the operator from mine debris under the working device. This kind of operator movement provides greater operator safety than when moving outside the trail lane. However, on top of that, the machine operator can still cause himself injury or sometimes even end up as a casualty from activated mines bounced from aside.

Escort vehicles and machines with direct control from its cabin, are tested for ballistic protection and safety of the machine operator. If the machine operator is located in the armoured cabin (of the demining machine or the escort vehicle) then he is protected by the machine design (protective armour thickness and protective window glass). During demining, cabin windows have to be closed. That means drivers and operators of the machine are protected inside the vehicle, so when they work, in theory, they needn't wear the PPE gear. However, machine operator and driver often leave the vehicle and come out to the mine-suspected surface, and therefore should wear certain PPE. The level of PPE protection pack can correspond to that of PPE worn by the machine operator moving behind the demining machine.

5.1.1 Machine Operator Risk with a Remote Controlled Machine

Experiences from using the machine are likewise important. According to CROMAC data [2], a 15 year experience shows that there were several machine operator casualties. Therefore, a special attention is to be given to a need of research and solving this issue. Often in practice, minimal PPE operator gear is used: protective vest and suitable light footwear (boots) and head (visor and light helmet). This type of machine operator equipment weighs usually 30–50 % less than the weight of standard PPE deminer gear. At the same time still the question remains about the operator protection and effects of AP fragmentation mines. For the time being, this issue is covered by measures of passive protection (operator being at a safe reach from the machine and moving along the silhouette of the machine).

This mode of remote control operation has displayed a certain amount of problems, from machine supervision, terrain configuration monitoring, quality of performance as well as monitoring area overlap in so that the operators are self-initiated in reducing the safety distance, risking their lives and health. By remotely controlling the demining machines the operators are exposed to increased risks and danger from uncontrollable detonations and in that regard have to be protected from splinter hits coming from anti-personnel and anti-tank mines, furthermore from impulse noise and vibration shock impacts.

Active protection refers to PPE of a machine operator. In practice, the level of protection is limited with the capability of machine operator movement. Due to higher equipment weight and fast machine operator weariness, his moveability lapses. This is why less equipment is favoured.

Depending upon the degree of possible danger in the mine-suspected area, demining companies use levels of protection in line with the threat posed. Basic PPE machine operator equipment remotely controlling, includes:

5.1.2 Protective Vest, Helmet and Footwear

Vest should protect frontal upper body part, including sides, neck, shoulders and upper parts of hips with groins. Additional protection for back, arms and legs can be used by operator's choice, i.e. demining company.

Helmet with visor has to protect face and its sides, forehead and neck. It should be designed to cover the collar of protective vest or it can be set into collar. It should provide good visibility without restricting the view.

Protective footwear should be comfortable and protect against threats. It is recommended, if possible, to wear countermine boots, or army type boots.

5.2 Requirements for Personal Protective Equipment

Personal protective equipment of the machine operator grounds on the features of PPE deminer. However, owing to the need of higher operator mobility in relation to that of deminer's one can speak of a minimum necessary level of machine operator equipment according to the zone of risk.

CWA 15756 quotes methods for testing, evaluation and accepting PPE for protection against AP blast mines. Excluded are fragmentation mines. PMN is chosen as a representative mine for this category of AP blast mines (240 grams of TNT). Most of other AP blast mines possess less explosive content (PMA-2 mine, 70 grams of TNT). The closer is the operator to the AP "cone of destruction", higher is the risk of injuries.

Table 5.1 Values of protective equipment V50

Personal protective equipment	Protective V50 value according to STANAG 2920, for 17 grain FPS
1. Protective suit	For chest, V50 ≥ 550 m/s For legs, arms V50 ≥ 450 m/s
2. Protective helmet	For helmet, V50 ≥ 550 m/s For visor, V50 ≥ 450 m/s
3. Protective vest	For waistcoat, V50 ≥ 550 m/s For groin V50 ≥ 450 m/s

[1] [2] [3] *Safety & occupational health—Personal protective equipment* guidelines, basic PPE characteristics are provided. Ballistic protection is evaluated based on V50 ($\mathbf{V_{50}}$) parameters, in accordance with STANAG 2920 [4]. In accordance with these standards an Ordinance on technical requirements and Conformity assessment is formed [5] that explains the requirements and testing of PPE in humanitarian demining.

5.2.1 Definition of V50 Value

V50 ballistic limit velocity for a material is defined as velocity at which the probability of penetration of the particular projectiles is exactly 0.5. Using the Up and Down firing method, the first round shall be loaded with the amount of propellant calculated to give a projectile a velocity equivalent to the estimated V50 ballistic limit of armour. After the defined number of projectiles has been fired, V50 is calculated as the mean velocity recorded for fair impacts, consisting of three highest velocities for partial penetration and three lowest velocities for complete penetration, providing that all six velocities fall within range of 40 m/s. Requirements providing very high protection for personal protective equipment in humanitarian demining on the basis of V50 values are given in Table 5.1.

Suit includes vest, sleeves, trousers. Suit has to be functional for all sizes (S, M, L, XL and XXL), so it will not deteriorate deminer's mobility and freedom of movement. Suit weight depends on size, but it should be less than 10 kg. Additionally, helmet weight with visor depends on size and type, and is

[1] Methods for determining the provided level of protection is defined in the IMAS in terms of blast and fragmentation effects as follows:

[2] Safety& occupationalhealth be capable of protecting against the **blast effects of 240 g of TNT** at stand-off distances, for each item of PPE, appropriate to the activity performed in accordance with SOPs.

[3] -ballistic body armour with a STANAG 2920 V50 rating (dry) of **450 m/s for 1.102 g fragments.** (Such tests for ballistic protection do not realistically replicate mine effects, but will continue to be used until an accepted alternative is developed as an international standard)".

Table 5.2 Helmet and visor characteristics (*Source* Ref. [7])

Characteristics	Helmet	Visor
Ballistic protection	V50 = 550/600/620 m/s, tested according to STANAG 2920 (same ballistic protection all over the surface)	V50 = 630 m/s
Material	*Aramide*	Fully laminated acrylic-polycarbonate ballistic visors
Special cover	Helmet is covered with special ABS-film	
Suspension	System with excellent shock absorption	
Chinstrap	Three point links width quick-release catch	
Nominal dimensions		150 mm
Mass	1.30 kg (small) 1.40 kg (medium) 1.55 kg (large)	cca 1.00 kg

1.0–1.5 kg. Standard protection for the visor is V50 $\geq$ 450 m/s against fragments or V50 $\geq$ 240 m/s protection against blast type mines.

5.2.2 Helmet and Visor Characteristics

It has been confirmed that helmets made of composite materials provide better ballistic protection in comparison to steel helmet of same mass. For the mass of 1.35 kg, helmet made of ballistic nylon has V_{50} around 450 m/s, and V_{50} for steel helmet is around 300 m/s. For *aramide fibres* (*kevlar, twaron*), for the same weight, V_{50} exceeds 600 m/s. Modern helmets, when bullet or fragment penetrates, provide selective resistance regarding their design and arrangement of layers (Table 5.2). At the begging, bullet or fragment penetrates through very hard layer, deforming in a shape like mushroom. After that, layers of different components, which fragment hits, are gradually stopping the fragment.

Helmet mass depends on its size: small, medium, large. Helmet mass excludes visor. Visor fits select PASGT helmets to provide enhanced demining protection against fragments from blast-type AP mines, Fig. 5.1.

Features

- Adjustable band fits select PASGT helmets
- Visor can be locked in the lowered position during operations or in the raised position during rest periods

Fig. 5.1 Helmet, BK3&9 V-150, PASGT with visor, 2.3 kg (*Source* Ref. [7])

- Visor integrates with chest plate to provide continuous frontal protection over the upper body, neck and head
- Easily attached to or removed from the helmet.

5.2.3 Protective Vest Characteristics

The protective vest is required to ensure total frontal protection from the fragments and impact wave of AP blast mines as well as the maximum protection from fragmentation mines in the full 360° circle of danger. The vest is also to provide all human movements, in the walking, kneeling and lying mode. Vest is made of *Aramide fibres*, improved with compact armour inserts that provide required ballistic protection level for vital body parts against AP mine fragments, and cushions are reducing friction, i.e. increasing mobility. Absorptive spine shield is of anthropometric design. Ballistic inserts for vest are respectively closed up, and with girder, made of nylon material, can be easily taken off for checkout and maintenance. Pockets and attachments on vest enable carrying different equipment, tools and radio-communication devices (Fig. 5.2) [8].

5.2.4 Mine Protective Boots and Covers

For reduction of AP mine influence on human body and legs, mine protective boots with covers are designed. Boot design includes shaped boot base (sole), vulcanized "sandwich" design in between of V-shape of external boot base and

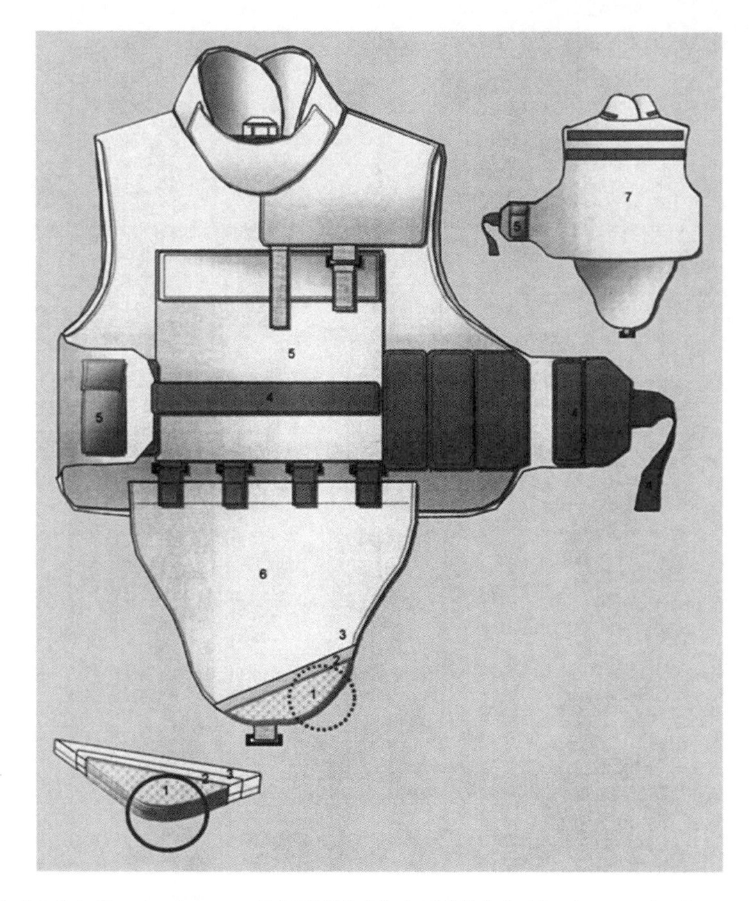

Fig. 5.2 Model of protective vest, BOROVO, Model B1/B2, 3.4 kg *1* protective layers (*aramide* PPTA fabric), *2* protective layers cover (hydrophobic), *3* vest outer layer (hydrophobic), *4 velcro* tape, *5* pocket, *6* groin protection, *7* back protection

multilayer *kevlar* inserts, which are compressed as honeycomb. Its purpose is to direct and to reduce AP mine explosion effects. Explosion pressure spreads to the sides of external boot base. Boots can be worn individually or combined with covers. Cover has, as an addition, upper multilayer *kevlar* protection, and combined with boots provides maximum protection. Slippers can be used in combination with other boots or shoes.

5.2.5 Acceptance Criterions

Personal protective equipment is tested according to criterions:

- ballistic testing against secondary fragments,

- testing blast effects on PPE system (using a dummy),
- testing ergonomic suitability of PPE.

Testing the level of protection from primary fragments is conducted in accordance with STANAG 2920 [4]. Behaviour of secondary fragments quite varies from the primary fragmentation. Therefore the testing takes place after the STANAG 2920 modifications in (a), (b), (c) and (d). The value of V50 is 1000 m/s. This value of V50 applies for materials of fibre type like *aramide* and *polycarbonate*. Other components of armour concerning various materials can amount to a different value of V50 for the same level of protection.

The scope of testing the PPE system when affected by the blast, according to CWA 15756, is to verify how various parts of PPE act together as protection and the integrity of equipment after the blast (blast test). Blast testing is performed on a testing male dummy Hybrid III by means of a device that enables precise dummy positioning in a certain pose. Mine simulator is filled with an explosive with the equivalent of an AP mine. Testing of ergonomic suitability is performed on the basis of talk with deminers and machine operators. Ergonomic suitability testing may be conducted by the demining company.

Conclusions

In lacking the escort vehicles of MPV type, a provisional solution option is suggested: a defence shield, a defence shield (protective door) on-board a multi-functional terrain vehicle, or a mobile protection cabin. Such devices are to provide protection from mine level threats in humanitarian demining MTL-02 (anti-personnel fragmentation mine and small UXO). The level of such protection matches STANAG 4569 [6], Level I and Level II (distance 30 m) or at least EN 1063. The lower level of protection is MTL-01 (anti-personnel fragmentation mine) or B2, B3 (EN 1063). A temporary solution to the safety of a machine operator might be a defence shield, protection cabin (mobile and on vehicle) or a protective wall on—board the terrain vehicle, that meet the requests for anti-mine protection MTL-01 [2].

References

1. Personal protective equipment (2007) CWA 15756, Test and evaluation, CEN, Humanitarian mine action, Brussels.
2. Mikulic D, Ban T (2012) Safety of demining machine operators, Book of Papers, International Symposium Humanitarian Demining, Sibenik.Protective helmet (2012) Power protection, Catalogue, Sestan Buch Ltd, Prelog.
3. Personal Protective Equipment (2001) IMAS 10.30, Edition 1, UNMAS, New York.
4. Ballistic Test Method for Personal Armour Materials and Combat Clothing (2003), STANAG 2920, Edition 7, Annex B, MAS, NATO, Brussels.
5. Ordinance on Technical Requirements and Conformity Assessment of Devices and Equipment Used for Humanitarian Demining (2007), National Gazette no. 53/07.

6. Protection Levels for occupants of Logistic and Light Armoured Vehicles (2004) STANAG 4569 (Edition 1) Military Agency for Standardization (MAS), NATO, Brusseles.
7. Protective vest against shrapnel/fragments (2012) Catalogue, Borovo-Gumitrade Ltd, Vukovar.
8. Protective vest against shrapnel/fragments B1 (2012) Catalogue, CHD—Cluster for humanitarian demining, Zagreb

Chapter 6
Test and Evaluation of Demining Machines

During CEN workshops, international specifications for testing and evaluation of demining machines were proposed. The main objective is that testing performed in accordance with these specifications should be accepted worldwide, which will provide guidance for demining machines development and comparison, requiring that machine testing is performed in specified and repeatable conditions. Former criteria for machine development were very different, leading to different engineering approach to machine development, testing and inadequate use. Nowadays view on demining machines role is completely altered in their favor.

Specifications were determined to ensure required performance and safety for people and machines in operation, and proper level of reliability and suitability, fulfilling operational requirements. CEN Workshop Agreement—CWA 15044:2004 on specifications favors development of new machines and demining methods providing easier comparison and evaluation of machines, and will improve efficiency of demining programs. Revision of this document was delivered on 30 June 2009, as CWA 15044:2009 [1]. Benefits from these specifications are acceptable worldwide. On basis of CWA a Croatian standard HRN 1142:2009 [2] was compiled, determining requirements and procedures for testing and evaluation of demining machines as well as their acceptability in humanitarian demining. Steps in demining machines evaluation are shown in Fig. 6.1. It is important, and should be emphasized, that this standard should be considered and applied by designers and end users.

6.1 Criteria for Evaluation of Demining Machines

The aim of HRN standard's specifications is to create industry-accepted criteria for the testing, evaluation, and acceptance of mechanical demining equipment. The standard [2] provides technical criteria for the following tests:

D. Mikulic, *Design of Demining Machines*,
DOI: 10.1007/978-1-4471-4504-2_6,

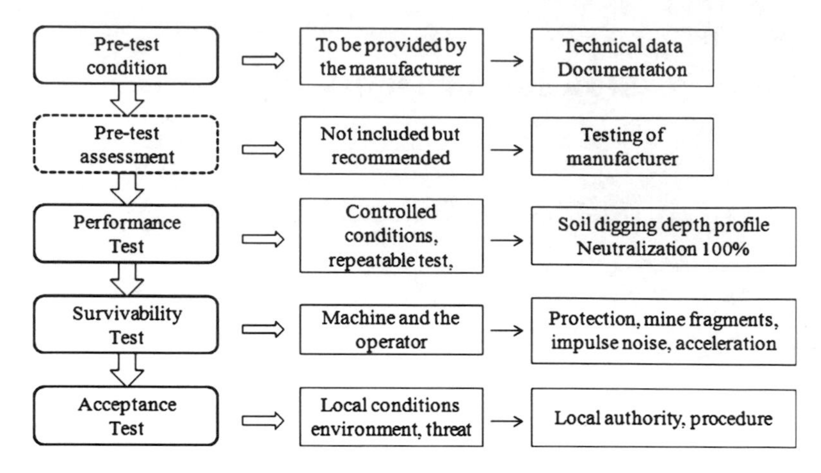

Fig. 6.1 Steps in demining machines evaluation

Performance test
Test to determine if the machine is capable for operation in a mine polluted environment under comparable and repeatable conditions and to evaluate the manufacturer's specifications.

Survivability test
Test of the effects of explosive forces on the machine and operators. The explosive force used is based on the threat level that the machine is designed for.

Acceptance test
Test to ensure that the machine is capable to work in the environment where it is intended to be used. The criteria should provide guidelines for local authorities when accrediting machines.

Basic test requirements for demining machines

- *Light* machines shall have only the capacity to neutralize AP mines but shall be tested against 8 kg explosives for survivability.
- *Medium* machines shall have the capacity to neutralize both AP and AT mines.
- *Heavy* machines shall have the capacity to neutralize both AP and AT mines.

Implementation of the criteria

Testing of machine performance is necessary for mine clearing and is performed in the different conditions and with various mine types. For survivability testing of the machine and explosion impact on the operator, different levels of protection and crew safety standards were proposed accordance to threats. For survivability testing of the machine in real conditions and soil, operational values are defined, and as a confirmation of this method, additional testing using conventional demining techniques should be performed.

Implementation of criteria provides advantages in machine development and evaluation. Present aspects of machine role in demining are completely changed in machine's favour. Standard for machine test will, as a result, provide innovations of mechanical technologies and competitiveness on market. The goal is simple: countries are developing capabilities to fulfil machine research and development tasks for specific needs according to common criteria. This provides development of resources that enables *own choice* in decision making process when demining is concerned. Additionally, it is bringing own strategy identity and role of technology in mechanized demining process.

6.2 Demining Machine Performance

Performance indicates the capabilities of a demining machine within specific working conditions, which are testable on a test range. The demining machine performance comprises the following indicators:

- Soil digging depth profile
- Reliability of mine neutralization (by destruction or activation)
- Soil digging speed

To achieve a quality soil digging depth profile, utilization of the machine's speed and rotation of the working tool is very important. For machine use, it is necessary to adjust machine forward speed v and flail rpm n, and to determine the technological hammer tool shear (hammer hit interval) S, with minimal side overlap. Graphic interpretation of soil digging density is called a mine clearance diagram. Practically, demining machines need to provide a quality tool penetration profile, i.e. soil digging depth profile of a working tool (flail, tiller), Fig. 6.2.

6.2.1 Soil Digging Depth Profile

To evaluate the penetration profile, sections of 4 mm fibreboard are put into the across the test lane of the machine, buried up to 15 cm below the maximum depth. The width of the fibreboard needs to be at least 10 % greater than the width of the digging tool. Joining several sections to achieve required width is acceptable. A minimum of three fibreboards are used, one in front of targets, one within targets, and one after targets. Measurement of soil digging depth is done on test lane, equipped with three fibreboards of 40 cm width, which are levelled with surface of compacted soil. Machine movement speed (without changing machine

Fig. 6.2 Soil digging depth profile of a flail machine

movement speed on test lane of 50 m length), provides fibreboards to be taken out and size and continuity of fibreboards damages are measured. At the same width as working tool width, fibreboards should have continuous damages, without any undamaged parts in marked depth. Undamaged fibreboards parts point to asymmetry in operation of hammer, i.e. engine power required for working tool operation is not adequate, and for this reason, undamaged AT mines can be found.

Acceptability of the soil digging depth profile (acceptance test) is defined by the standard [2] and ordinance, while the accredited institution CROMAC CTDT determines the organization and procedures, as well as executes testing.

Reliability of mine neutralization is evaluated by the probability that the working tool will neutralize mines which are placed in its operational area. This probability is theoretical expected ratio between number of mines that were removed (neutralized), and total number of mines laid in particular area. For that reason, it can be called percentage of clearance of tested machine. Probability of 95 % of clearance percentage means that 95 mines (of 100) can be found, and other 5 will not be found (risk is 5 %). According to standard HRN [2], all laid AP and AT mines have to be neutralized (100 %).

Machine has to maintain required density and digging depth, and reduction of forward speed if rotor rpm drops significantly. Tool shear has to maintain constant: if machine forward speed is increased, than the rotor rpm shall increase, and if machine forward speed is reduced than rotor rotation speed shall be also reduced. In machine testing it is important to evaluate machine reliability of mine neutralization, i.e. mine clearance reliability. Based on those results, technical solutions of machine and working tool can be evaluated. Testing conditions have to be repeatable: mine types, laying depth and treated soil type. Additionally, it is possible to compare clearance reliability of machines in particular mine clearing conditions.

The main purpose is to test demining machine performance in humanitarian demining, in order to identify equipment that is safe, reliable and suitable for its purpose. Testing of mine clearance in controlled conditions verifies the capability of machine to neutralize mines up to different depths in different soil types. Testing should be done on testing lanes on three different soil configurations to different depths, from surface to maximal penetration depth (according to data provided by manufacturer).

Before every run the soil shall be cultivated, or otherwise loosened up, and then compacted to its original state again. The level of compactness is to be measured and recorded using 10 points randomly distributed along the lane. Level of soil compactness is estimated around 80 % of maximal value.

Performed testing and measurement on lanes also includes measurement of soil digging speed, providing a basis for estimation of machine efficiency.

6.3 Survivability and Protection

In demining machine development, machine operator safety is the most important factor. When mine is activated close to the machine, blast wave is created and is spread in all directions faster than speed of sound. This wave surrounds machine cockpit causing floor, roof and overall plating vibrations, creating overpressure inside the cockpit. Cockpit is not hermetically closed, and blast wave and explosion overpressure penetrate into the cockpit causing short and sudden pressure variation.

Blast wave and vibrations can cause injuries to the crew. Blast wave can hurt unprotected ears, but also internal organs that are filled with air. Vibrations and sudden position changes can cause complex injuries of foot and ankle if leaned on the floor, as well as spine injuries.

Crew protection can be categorized according to machine armour protection. Testing of mine explosion influence on machine provides data, based on which overall crew protection system can be evaluated. Goal of studying interactions and consequences of mine detonation on machine structure is to compile physical effects of mine explosions into models that will be basis for machine evaluation on explosion resistance and determination of safety level and safety requirements. Machine design consists of inner and external protection. External, armoured protection reduces impact of fragments energy. Inner protection is used for protection against vibration and noise. When describing certain injuries and protection, noise and vibration criteria are adopted.

Tests are focusing on two areas:

- Machine survivability—mine blast effect on the machine.
- Operator survivability—level of protection provided to operators subjected to the effects of blast.

Before machine testing, it is necessary to evaluate protection specifications:

- Materials used (types, thickness, certificates, etc.);
- Design principles (blast deflection, distances, etc.);

- Construction quality (fittings and controls access, welds, etc.);
- Safety principles (such as exits, fire suppression, etc.).

Explosion effect on working device is measured in controlled environment. In order to avoid unnecessary damages, smallest mine is laid first. Mine is laid under the centre of the tool. Depending on the results, second mine of the same explosion force is detonated at the end of tool. All mines are detonated remotely.

Operator survivability

The aim is to verify the survivability of the crew after AT mine detonation. The following effects will be measured and evaluated:

- Overpressure in internal organs (ear);
- Acceleration (feet and spine);
- Displacement of operator.

The mine is placed in the area deemed most likely to have maximum effect on the operator (worst case scenario) e.g. under the working tool, wheel or track bogey closest to crew compartment. Mine placement is in direct contact with the target area. All mines will be initiated remotely.

6.3.1 Operator Protection Against Mine Fragments

Countermine protection level for the crew of logistics vehicles are defined with STANAG 4569 Annex B. Protection levels (I, II, III, IV) depend on quantity of explosive and position where is activated—under the tires or under the centre of the vehicle.

Protection system for operator and crew is of modular design. Floor board countermine protection is designed according to the intensity of "cone of destruction", intensity of reflected pressure and "side on" pressure. In case of direct AT mine activation under the machine tracks, machine will be stopped due to detonation products pressure.

For protection against PROM-1 AP bouncing-fragmentation mine, ballistic materials are used. Most of the logistic vehicles use ballistic armour plates High Hardness Steel (HHS), of 480–540 HB (Brinell hardness). Modular metal armour is mounted at certain angle to provide better protection. For armour thickness of 4 mm, 6 mm and 8 mm, vehicle protection is assured by positioning the plates under the angle of 25°–35°. Well-known materials used for armour design are steel *Armox* 500 plates, of 480–540 HB, thickness of 4–6–8 mm, or titan, which provides same protection level, but is lighter.

Windshields windows and other glass surfaces are usually made from so called ballistic glass—multi-layer glass, consisting of several glass layers, with protective membranes in between, in order to prevent penetration of fragments into the vehicle. For example, fragments from PROM-1, bouncing AP mine, are more dangerous for machine operator than infantry bullets. Mine can be activated under demining tool, under machine or very close to the machine. Machine operator should be protected against AP and AT mine fragments and blown off machine parts.

Demining machine protection concept:

Basic machine

- modular armour design
- armouring of easily affected parts of working arm
- armouring of easily affected parts of chassis

Working tools

- armoured hydro components,
- flail
- tiller
- vegetation cutter
- tongs and other special tools

Cockpit

- for demining—mono-block cockpit,
- visibility 360°, roof safety escape hatch, pressure valve
- multi-layer safety glass

Armour materiel

- *Armox* 500, 480–540 HB, thickness 4–6–8 mm
- titan/hardox/kevlar/multi-layer glass

Operator's cockpit safety (countermine protection)

- explosion of AT mine with 8 kg TNT, under the working tool:
- impulse noise protection (max 140 dB), and
- acceleration protection (less than 15 g at anti-mine seat)

Mine threat level (MTL)

- MTL-01 (anti-personnel blast)
- MTL-02 (anti-personnel fragmentation mine and small UXO)
- MTL-03 A/B (blast anti-tank mines under wheel and hull)

Equivalent to MTL 01/02/03, countermine protection:

STANAG 4569, Anex B, countermine protection:

- level I AP explosive devices
- level II 6 kg TNT
- level III 8 kg TNT

6.3.2 Protection Against Impulse Noise

6.3.2.1 Tolerance Level

Organs filled with air react to variation of air pressure. Ear is one of the most sensitive organs filled with air, and reacts to the smallest variation in pressure. Blast wave intensity (i.e. impulse noise, dB) is based on sound pressure (overpressure). Highest noise level should not exceed allowed values for hearing organs. For overpressure higher than 200 Pa (140 dB), protection equipment must be used (ear protectors, safety helmet, etc.). Conditions for operator safety in the cockpit after the mine explosion are based on "no injuries at all" principle, and operator should be able to continue to operate the machine (unless machine is destroyed or otherwise disabled) or to continue operation in new machine.

Determination of maximal overpressure

Maximum value of overpressure is the highest value achieved in any moment of mine explosion. Maximum overpressure may be expressed in dB (decibel), and conversion is done as follows: maximum overpressure (dB) = 154 + 20 log (maximum overpressure kPa).

Duration of sound overpressure impulse from the explosion of one TMA-3 AT mine (6.5 kg TNT) is 14.7 ms, while noise level peak is 153 dB. Peak value of explosion pressure is 4.8 kPa. Person exposed to this noise, wearing ear protectors which decrease noise level for 25 dB, can bear increased number of explosions without any damage to hearing sense, but allowed tolerance values in daily operations cannot be exceeded. According to mathematic calculation method used in military minefield penetration, daily tolerability limit is 11 AT mines.

To prevent undesirable psycho-physiological reactions, or to prevent temporary or permanent loss of hearing sense, harmful sound pressure effects on hearing sense has to be prevented by the use of protective equipment (ear protectors, ear muffs, protective helmet, safety vent for cockpit overpressure, etc.).

In order to achieve better attenuation of noise and improve personnel safety during the machine detection, operator cockpit interior is covered with sound absorptive materials. Beside interior, external machine parts have certain effects on noise level. Armoured protection, thickness of armoured plates of such machines protects operator against fragments and decreases noise level within machine cockpit.

Isolation of cockpit interior against noise includes isolation with sound barriers and isolation with absorptive materials. Foundation between isolation materials and surface is mostly based on natural gum or synthetic rubber covered with sticky resin, oil and antioxidants. Beside its use as sticky resource for isolation, it provides good adherence on surface that needs to be isolated.

Beside cockpit isolation, operator is liable to use protective equipment such as helmet, ear plugs and ear muffs. Use of protective equipment is mandatory for safety reasons, because it is not possible to precisely determine quality of cockpit sound isolation. Values can be determined with precise testing, and depend on value of peak overpressure, machine and cockpit shape, etc.

6.3.3 Protection Against Vibrations

High exposure to vibration impacts increases the risk of spine injuries and pain in spine lumbar area. Extremely high values of impact, such as driving on extremely rugged terrain or mine explosion under the vehicle, can even cause spine fracture. Additionally, long exposure to low impact values causes degeneration of spinal discus, which leads to constant pain.

Beside above mentioned possibilities of vibration impact effects on humans, scientists developed different methods to measure exposure to vibrations. One of the most competent methods used is ISO 2631 method, revised in 1997, which describes mechanical vibrations and evaluation of human body exposure to vibrations Whole-Body Vibration (WBV). Standard requires the use of new method for evaluation of exposure to vibrations, called Vibration Dose Value (VDV). This method is used when during exposure to approximately constant vibration values, vibration impacts appear at the same time. UK standard also prescribes VDV—when one speaks about exposure to vibrations with multiple impacts.

Operator seat

Seats used on some demining machines have suspension that is using air bags and passive silencers to isolate operator from vibrations between 4 and 8 Hz, because human body is most sensible to vibrations in this range. When exposed to vibration impact these seats are becoming too soft and suspension is moved to its utter end position. One of the solutions is to mount the rubber limiters on the seat, which will partially absorb the impact. That does not provide adequate protection for operator—when impact affects the seat, seat is quickly stopped resulting in twitches or jerks and possible injuries. By adding springs and silencers or replacing the existing ones with more rigid ones, it is possible to prevent a contact between the seat and rubber limiters during the vibration impact, but operator isolation in the most sensitive frequency area will be lost.

Requirements for seats (operator stability, reduction of vibrations, etc.) have to be considered for design calculations or at seat selection, in accordance with limitations for protection against vibrations.

Seat has to provide stable position to operator, and designed in accordance with ergonomic principles. Seat has to be designed in order to reduce to the

Table 6.1 Crew safety—no injury levels for the ear, foot/ankle and spine

Physical effect	Body part	Level
Pressure	Ear	<W-curve (140 dB), no protectors required >W-curve but < Z-curve, protectors required. >Z-curve is not allowed
Shock/ acceleration	Foot/ankle	Average acceleration < 20 g or max velocity change < 3 m/s
	Spine	Average acceleration < 15 g or max velocity change < 4.5 m/s DRI ≤ 16 Dynamic response index

(*Source* Ref. [3])

lowest possible level vibrations transferred to operator. Seats mount has to endure required level of stress during operation and in case if vehicle rolls over. Seat has to be equipped with safety belt or an adequate device that is preventing unnecessary and undesirable operator movements and which will keep him in a seat in case of vibration impact, not limiting operator ability to operate the machine, or affecting proper operation of seat suspension. Machine has to be designed to prevent cockpit deformations in case if machine rolls over, reducing the possibility of injuring the operator. Seat belts or other adequate equipment have to keep operator in the seat in case if machine rolls over.

Machine cockpit is a subject to operator protection standards, such as SAE or ISO. Cockpit design has to provide protection if machine rolls over Roll Over Protective Structures (ROPS) and against falling objects Falling Objects Protective Structures (FOPS), according to SAE J394, ISO 3471, SAE J231, and ISO 3449 criteria.

Due to mine explosion, chassis and the rest of the machine are vibrating with amplitude and frequency depending on explosion intensity (type of explosive device and position where it was activated), machine suspension type, cockpit suspension (if there is one) and operator seat suspension. Vibration shape is affected by machine geometrical and physical characteristics, such as chassis shape, floor thickness, material used, hull, etc.

The acceptable injury level for the occupants is set to "no injury at all" implying that after a detonation of a mine under the clearing device the occupants will be able to continue their mission in the vehicle, if undamaged or exit and operate another vehicle, as is shown per Table 6.1 [3].

Due to large changes in vertical acceleration in short time interval, it is most important that operator sits properly. Proper sitting reduces the load on spine and whole musculature, which causes fatigue and directly effects operator concentration if operating machine for longer period. During the explosion, it is important that the whole body, particularly the spine, is in ideal position, in order to reduce stress and to minimize the possibility of injury.

Most people do not pay proper attention to sitting position, and if they do, during longer working period they tend to loosen up. Some of the causes are badly designed seat, unsatisfactory ergonomics, and bad design of working environment. All of these factors are also causing fatigue. Seats have to be designed to keep the body in ideal sitting position without operator even thinking about his sitting, i.e. when occupying his seat operator will immediately have proper body position.

The seat must be convex in the lumbar part, maintaining a natural convex spine curve, reducing its stress from body weight, as well as fatigue and pain. Such a seat back is not pressing the spine, but keeps it in the proper position. For the proper spine position, beside the back of the seat, position of pelvic bone is also very important. It is not possible to achieve regular spine curve if both conditions are not fulfilled (shape of the seat back and pelvic bone position, i.e. angle of sitting position).

Body weight is, through pelvic bone, leaned on foundation, but mass centre is behind the "sitting" bones. During high acceleration in short time period, pelvic bone is moved down and back, causing the pressure on the lumbar part, i.e. compressing the frontal part of spine disc and causing pain. Such an improper spine position reduces injury resistance for 50 % during vertical acceleration.

All assumptions for seat design and selection should comply with one condition: keep the body in proper position while seated.

6.4 *Cerovac* Test Range

Centre for Testing, Development and Training/CROMAC–CTDT[1] is established by Croatian Government Agreement with the goal of scientific-research and educational activities in countermine actions on income basis. Basic centre's tasks are testing of methods, technologies and equipment for humanitarian demining (procedures, standards, detectors, dogs, demining machines, equipment, etc.), maintenance of existing and establishment of new test ranges, development of demining standards, scientific and expert cooperation with national and international institutions. In accordance with humanitarian demining standards, machines and other equipment is being tested first. In order to ensure rational use of test ranges in Croatia, *Cerovac* test range is suitable and has a dual use.

Demining companies are equipped and are working with modern equipment and technologies for humanitarian demining. Technology for humanitarian demining can be used for military purposes in peace keeping and PfP operations,

[1] CROMAC–CTDT. In collaboration with the Croatian Standardization Institute national standards for testing and licensing of tested mechanical equipment, dogs and new detection methods are performed. With the CROMAC support, the Centre strives to cross national borders with the aim of more efficiently implementing mine action in other countries of the region and beyond.

and it is necessary to achieve a compatibility of humanitarian and military demining within Croatian Forces. For this purpose, it has been determined a dual role test range, for example Swedish SWEDEC, Canadian CCMAT.

Estimate of interests

- Long-term use of *Cerovac* test range for testing, simulation and certification of equipment, and for education of all parties in humanitarian demining process;
- Training of engineer corps in accordance with standards for humanitarian demining for peace keeping and PfP operations;
- Settlement of test range for machines, dogs, detectors and mines testing, build up of facilities for testing and verification of equipment characteristics, based on investment and donations;
- Certification of test range in accordance with ISO standards;
- Licensing of officers and technical personnel/license/;
- Education of PfP and NATO member countries for humanitarian demining.

On test range, demining machines, mine detectors and detection dogs are tested, and deminer and other personnel training is performed, as well as research and development of new technologies in controlled and natural environment.

Test site for testing of demining machines was created by CROMAC–CTDT on the part of an existing military range.

Test range *Cerovac* area: 55 ha

Test range comprises:

Demining machines testing area
Detection dog and dog handler testing area
All detector types testing area
Personnel training area
Area for research and development of new demining technologies and methods

6.4.1 Demining Machines Testing Area

Demining machines testing area is designated for testing of all demining machine categories and annual evaluation of machine characteristics, Fig. 6.3.

At test range, working areas are determined for:

- Demining machine clearance quality testing
- Demining machine testing on AP and AT mine detonations
- Machine vegetation removal testing

Fig. 6.3 Demining machine testing lane (Reproduced with permission from CROMAC–CTDT)

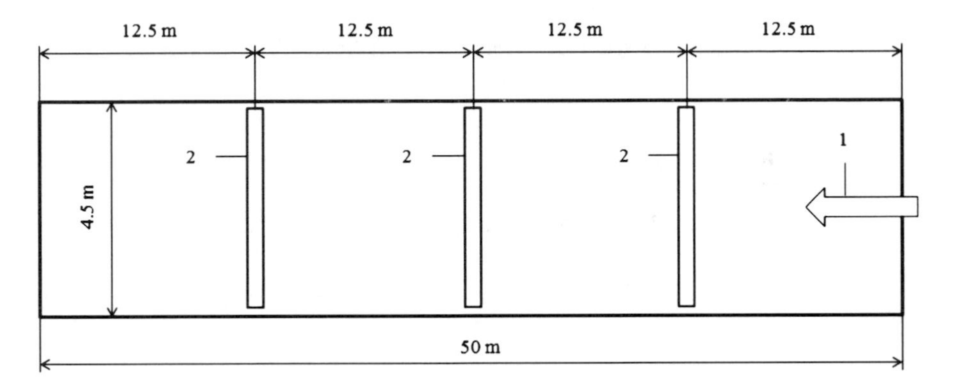

Fig. 6.4 Scheme of testing lane for demining machines *1* machine movement, *2* fibreboards (Reproduced with permission from Ref. [2])

6.4.1.1 Working Area for Demining Machine Clearance Quality Testing

A scheme of the testing lane for demining machines is shown on Fig. 6.4, while a scheme of the testing lane for excavators is given in Fig. 6.5. Three types of testing lane are available with different soil type and granulation, *Gravel, Sand and Ground (Topsoil),* Fig. 6.6.

Performed testing and measurement include:

- Soil digging depth profile (3 different soil types)
- Reliability of mine neutralization
- Soil digging speed (based on that efficiency is calculated and evaluated)

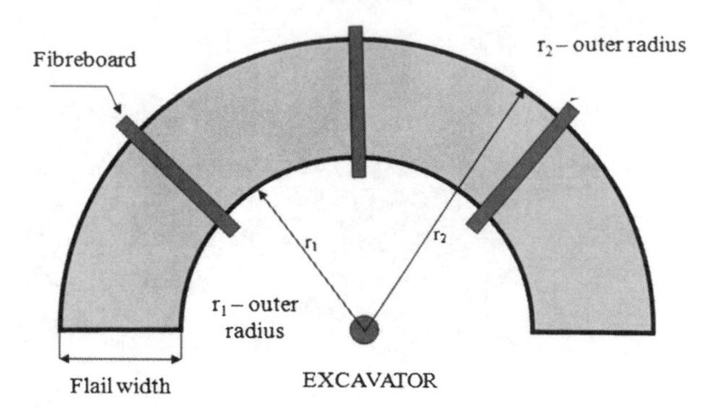

Fig. 6.5 Scheme of testing lane for excavators (Reproduced with permission from Ref. [2])

6.4.1.2 Working Area for Demining Machine Testing on AP and AT Mine Detonations

When demining machine testing is performed, all AP and AT mine types are used, as well as their surrogates.

Goal:

- To get results on capabilities and reliability of neutralization of all AP mines embedded down to 20 cm,
- To get results for machine and working tool endurance on all AP mine detonations,
- To get results on capabilities and reliability of neutralization of all AT mines,
- To get results on machine and working tool endurance on all AT mine detonations,
- To test crew safety within machine cockpit—noise and vibration measurement in machine cockpit,
- To test countermine protection for an escort vehicle.

6.4.1.3 Working Area for Machine Vegetation Removal Testing

Different types of vegetation can be identified (low, medium, high vegetation and single threes) on the soil of different inclination. Machines are tested in these conditions.

6.4.2 Detection Dog and Dog Handler Testing Area

Detection dog and dog handler testing area is physically separated from other areas, due to special conditions for the use of dogs and embedded mines. At test range, following activities are performed:

- Obedience exercise and training,
- Fitness exercise and training,

- Explosive detection exercise and training.

The licensing testing is also performed:
Licensing area 1—terrain demining
Licensing area 2—tripwire detection
Licensing area 3—secondary demining, verification after machine demining

Exercise and training area

Area for dogs exercise, training or warm up is available. Area has a number of boxes or search lanes with the same testing objects, as on the testing fields. Area is set up outside the testing area.

Licensing area 1—Terrain demining

Test range consist 80 test field boxes of 10 × 10 m in size where mines and explosive ordnance of different types can be found in particular pattern.

When preparing test range, (testing of licensing area 1), the following rules apply:

- Position of each box must be mapped precisely,
- All corners are marked with poles—metal marker. Pole is fixed in the soil, and its top is at the surface level. At least one corner marker is mapped precisely,
- All box sides are marked with tape (or similar), before test objects are placed,
- Exact locations of all test objects and metal markers are registered. Boxes should be of square shape,
- Testing area should have one or more identification landmarks. Distance and compass direction should be calculated from one corner marker for each box to the landmark. This simplifies preparation and orientation of testing area maps and locations of corner markers, which are sometimes difficult to locate during to rainy or winter period,
- Distance between boxes should be at least 3 m, if possible 5 m or more,
- Distance between mine in each box should be at least 3 m,
- Testing objects are embedded down to different depths; they can be on surface—camouflaged, or at maximum depth of 20 cm.

Licensing area 2—Tripwire detection

Test range consist 12 test field boxes of 10 × 10 m in size. Goal is to evaluate dog capability to detect small number of tripwires that cannot be seen by dog or dog handler, and are placed similar as in real conditions.

Following rules apply:

- Tripwire testing is physically separated from licensing area 1,
- For testing purposes, camouflaged tripwires are used (covered with leaves, grass or other natural material),
- Boxes (10 × 10 m) are partitioned into squares of 2 × 2 m in size, to ease the measurement. In some boxes, tripwires are placed diagonally,
- Searching lane is 1 m wide and is placed centrally inside of 2 m squares,
- Position of each box is mapped precisely,

- All box corners are marked with tapes (or similar), before placing testing objects,
- Position of all tripwires is recorded,
- Distance between any part of tripwire and neighbouring tripwire is at least 3 m,
- Each tripwire placed within box is at least 2.5 m long.

Licensing area 3—secondary demining, verification after machine demining

Licensing area 3 is similar to Licensing area 1, except for testing preparations. For purpose of this testing, area of 42 m × 42 m is available.

6.4.3 All Detector Types Testing Area

Area enables testing of all detector types: metal detectors, GPR, IR detectors, and multi sensor detectors. Width and depth of testing lanes are specified so that soil on local test range (if different from the soil on test lane) could not influence detector during testing. Testing area, with embedded mines, is 1 m wide and is placed in the middle of the lane. Testing lane length will be determined based on number of targets in each lane (single types and quantity for multiple settings). For testing purposes there are 5 lanes, each 1.5 m wide, 0.5 m deep, and 30 m long. Each lane has the soil of different pedological and electromagnetic features.

Distance between testing lanes is 3 m.

In front of each lane, there is a space available for detector adjustment and calibration.

6.4.4 Personnel Training Area

On specified area, it is possible to conduct individual and team training, as well as exercises with different live and training mine-explosive ordnance.

Individual training

- Training on demining of all training AP and AT mines,
- Training on demining of all types of training mine-explosive ordnance,
- Training on use of metal detector for surface and in depth detection,
- Training on use of deminer personal equipment (all types and for all purposes),
- Demonstration of demining operations and procedures for all AP and AT mines.

Team training

- Organization of demining site,
- Training on demining operations and procedures on site (deminer, team leader, site manager, medical team, physical security, deminer surveillance)
- Demonstration and training on survey of mine suspicious area,
- Demonstration and training on medical assistance,
- Demonstration on use of methods in integrated connectivity on working site.

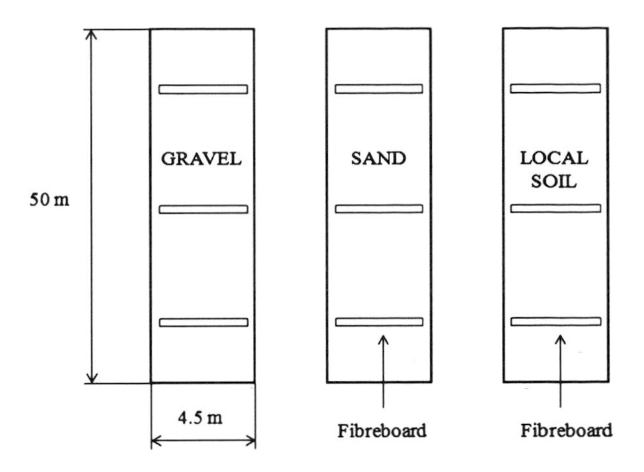

Fig. 6.6 Test lanes: gravel, sand, local soil

Research and development of new demining technologies and methods testing area

For research and testing purposes of new demining technologies and integrated connectivity, natural environment area of 25 ha in size is designated. Inclination of terrain is 3–10 %, vegetation height is 0.5–4.0 m.

6.5 Demining Machine Testing Process

Machine demining provides primarily deminer safety, and meeting the high quality standards in soil treatment removes mine threats. This includes soil digging at certain depth, vegetation removal and destruction of mine activation wires. Secondly, it is important to prepare surface for survey using dogs or detectors. Such standard operating procedure is basis for defining of demining machines testing process, Fig. 6.7.

Minimum of demining machine testing includes:

- New machines—full testing,
- Machines used—annual compliance testing

Demining machine testing process includes following activities:

(a) Preparation of documentation
(b) Testing of digging depth profile and efficiency
(C) Testing on AP mines
(d) Testing on AT mines
(e) Operator protection testing
(f) Testing of excavators
(g) Testing of an escort vehicle
(h) Annual compliance testing
(i) Additional testing

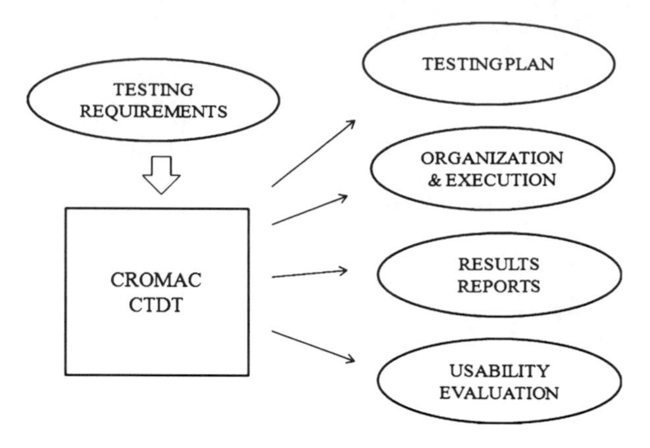

Fig. 6.7 Demining machine testing process

Each of above stated activity is integral part of unit interconnected, depending on machine, its technical characteristics and testing requirements.

6.5.1 Preparation of Documentation

Customer that requires demining machine testing may be authorized legal entity (demining company, or manufacturer). Machine testing requirement is submitted to CROMAC–CTDT, with enclosed following documentation:

- Machine technical data
- Manufacturer's certificate on material quality for countermine protection
- Development and field testing documentation
- Documentation on previous machine testing and operation.

Demining machine testing manager develops "Machine testing plan", which is presented to CROMAC-CTDT demining machine testing Committee. Committee can adopt proposed Plan, but may request changes and additions to the Plan. After Plan adoption, organization of testing and machine testing process may start.

6.5.2 Testing of Depth and Efficiency

Machine testing on test range is performed at the *Cerovac* test range, on soil types specially prepared and designed for testing (test lanes of 50 m length, 4.5 m width and 0.5 m depth), Fig. 6.7. On first test lane, *topsoil* cleared from stones and objects, in layers of 15 cm height was shot up, until desired layer thickness of 50 cm was gained.

On second test lane, *sand* was shot up, particle size of 0.075–20 mm, and 85 % of them less than 0.6 mm of size, regularly distributed. Sand was shot up in layers of 15 cm and compact, until desired layer thickness of 50 cm was gained.

On third test lane, *gravel* was shot up, particles size of 0.075–45 mm, and 10 % of them less than 0.4 mm of size, linearly distributed. Gravel was shot up in layers of 15 cm and compact, until desired layer thickness of 50 cm was gained.

On each test lane three fibreboards were set up (on the first third of length, in the middle of the lane and on the last third of lane), vertically in machine movement direction. Fibreboard thickness is 4 mm, length 4.00 m, and width 40 cm. Fibreboards are used for measuring the treating depth in different soil types. Depth, on which fibreboard is damaged, represents digging depth of certain soil type, i.e. toll penetration profile.

On each test lane, level of soil compactness is measured in order to provide the same (or similar) soil compactness level for each machine tested. This factor is important for machine testing, due to fact that each soil has different compactness, which provides different results of mine clearing (resistance, movement speed, digging depth).

On test lane, all machine categories are introduced (light, medium and heavy) and following parameters are tested: soil digging depth profile, and machine working efficiency

Soil digging depth profile is provided by measurement of damages of three fibreboards on each test lane (average), and soil treatment depth is provided for all three test lanes. Fibreboard damage width is measured on 20 different spots on same distance between the spots. Total number of measurement of soil treatment depth on one test lane is 60 measurements (20 on each fibreboard).

Machine efficiency on test lanes is gained based on machine movement speed and working tool width. Mathematical processing of these parameters provides possible machine working capacity in time unit (m^2/h). On different soil types, machine capacity is different, depending on compactness, treatment depth, machine type, engine power, working tool type, machine management and steering, etc. When soil treatment depth increases—soil treatment resistance increases too, which reduces machine movement speed. This way, machine working efficiency can be measured accurately and then be compared to theoretical efficiency diagram.

Gained results from machine testing on test lanes are highest possible values, due to ideal conditions. On test lanes there are no vegetation, no mines or explosive ordnance, machine movement is linear (no turning) and relatively short (50 m). On test lanes, machines can obtain best results due to following conditions and procedures:

- Same conditions for all machine types
- Repeatable procedure for all machine types
- Gained results are optimal (compared to real demining conditions)
- Procedure is simple and of short-time (up to 2 h), although preparations of test lanes takes significantly more time
- Possibility of comparison of results and their statistical processing

Mine type	Quantity
PMA-1A	5
PMA-2	5
PMA-3	5
PMR-2A	2
PROM-1	2
Total	19

Fig. 6.8 Types and quantity of AP mines for testing of demining machines (Reproduced with permission from CROMAC–CTDT)

6.5.3 Machine Testing on AP Mines

Types and quantity of AP mines for testing of demining machines are shown in Fig. 6.8. Mines used for testing were [2]: AP blast mine (PMA-1A, PMA-2 and PMA-3) are set linearly, at distance of 4 m and depth of 5, 10, 10, 15 and 20 cm, Fig. 6.9. The goal is to neutralize mines in one machine pass. Activated or crushed mine is considered as neutralized. Due to small quantity of explosive in above mentioned AP mines, damages on working tool or machine are not expected.

AP fragmentation mines (PMR-2A and PROM-1) are set and armed individually [2]. One PMR-2A mine is set in front of machine working tool at distance of 20 m, and tripwire is tight opposite to machine movement direction. At second PMR-2A mine, tripwire is tight vertically in relation to machine movement direction.

AP bouncing-fragmentation mines PROM-1 are set according to the scheme in Fig. 6.10. They are laid in front of machine working tool, one at a distance of 5 m

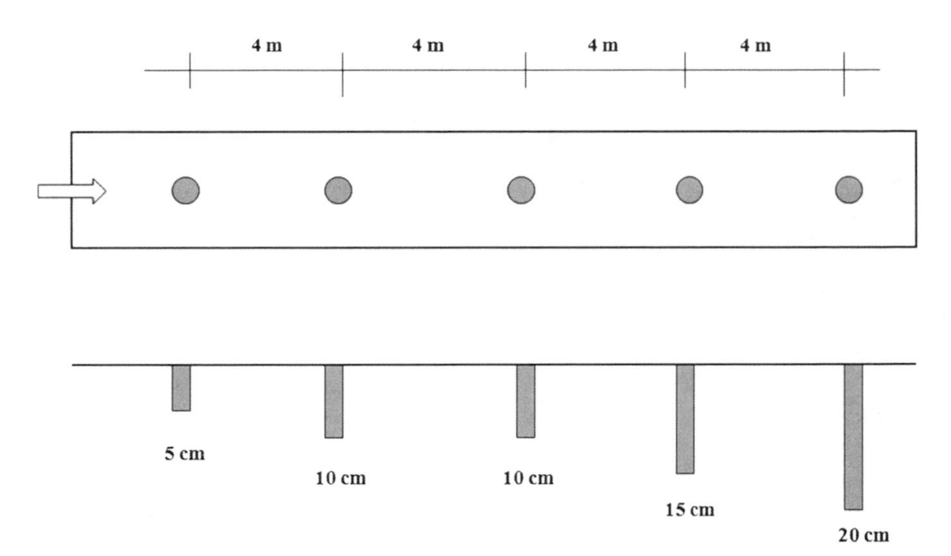

Fig. 6.9 Scheme of laying blast AP mines (*Source* CROMAC–CTDT)

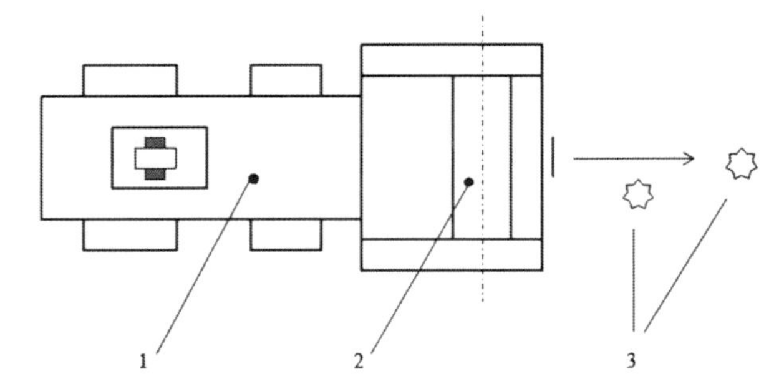

Fig. 6.10 Setting bouncing-fragmentation AP mines (PROM); *1* machine, *2* working tool, *3* AP mines (Reproduced with permission from Ref. [2])

pressure activated, and the other one at a distance of around 15 m with tripwire tight in the same direction as machine movement.

Due to higher explosive quantity at AP fragmentation mines, smaller machine damages can be expected, primarily on working tool. When machine is tested on AP mines, following results can be expected:

- all mines have to be neutralized (by crushing or activation)
- on working tool or machine, vital design parts have to undamaged; machine should continue operation.

When, after machine and working tool operation, undamaged mine is found, testing procedure is repeated, using the same number of mines, as well as types. If, after repeated testing procedure, undamaged mine is found, it is required that machine user performs a detailed checkout of machine and working tool, and requests for certain design adjustments on machine.

Mine type	Quantity	
Machine	Medium	Heavy
TMM-1	1	1
TMA-3	1	1
TMA-4	-	1
TMRP-6	2	2
Total	5	4

Fig. 6.11 Types and quantity of AT mines for testing of demining machines (Reproduced with permission from CROMAC–CTDT)

6.5.4 Machine Testing on AT Mines

On specially designed area, under safe conditions, machine testing on AT mines is performed. Types and quantity of AT mines for testing of demining machines are shown in Fig. 6.11. Considering machine type, following AT mines are used:

- Light machines and excavators are not tested on AT mines,
- Medium and heavy machines are tested on AT mines.

Remotely controlled medium machines, which are not equipped with operator's cockpit, are tested on AT mines.

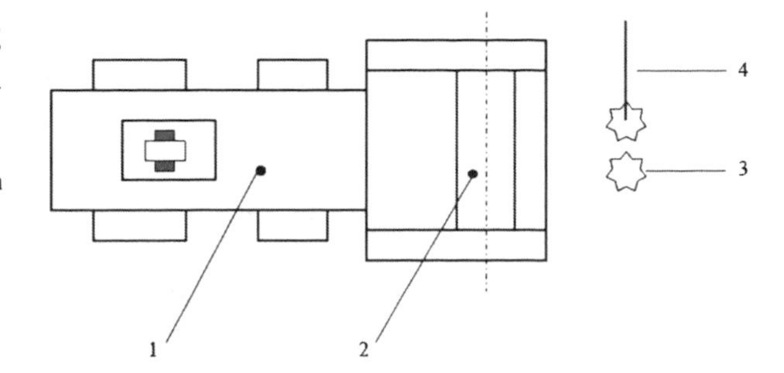

Fig. 6.12 Setting AT mines; *1* machine, *2* working tool, *3* AT mines (TMM-1 or TMA-3 or TMA-4), *4* AT mine TMRP-6 with antenna (Reproduced with permission from Ref. [2])

AT mines (TMM-1, TMA-3 I TMA-4), at machine testing, are set individually, in front of working tool at distance of 5 m and are embedded at depth of 10–12 cm, and armed with adequate fuse, Fig. 6.12.

TMRP-6 mine is set for pressure activation at a depth of 10–12 cm in front of the machine and working tool, at a distance of around 5 m, and the other one is activated by moving the lever. After AT mine activation—Fig. 6.13 on machine and working tool, vital parts of design should be undamaged; damages should be reparable on the site.

At machine testing on AT mines, following results are obtained:

- Machine influence on mine, and
- Mine explosion influence on machine.

AT mine can be neutralized by detonation or crushing. AT mine should not be lagged or rejected, i.e. should not be undamaged. When working tool activate AT mine, full or partial detonation can be achieved. Full detonation will be achieved when total quantity of mine explosive is activated. Partial detonation will be achieved when working tool crush the mine and some mine parts containing fuse are detached, and then fuse activation and the rest quantity of mine explosive. Partial detonation can be achieved at AT mines, which do not have body (e.g. TMA-3) or which have plastic body. With AT mines with metal body (e.g. TMM-1) there is no partial detonation.

Results of machine testing on AT mines point to the following:

- Expected AT mine destruction mode
- Expected consequences on the machine treated soil with AT mines,
- Expected damages on machine and working tool,
- Machine operator safety (in the cockpit or outside cockpit)
- Machine operator position
- Evaluation of need for spare parts (both working tool and machine)

6.5.5 Machine Operator Protection Testing

Medium and heavy machines may be equipped with cockpit, and machine steering can be done from the cockpit or outside the machine—remotely

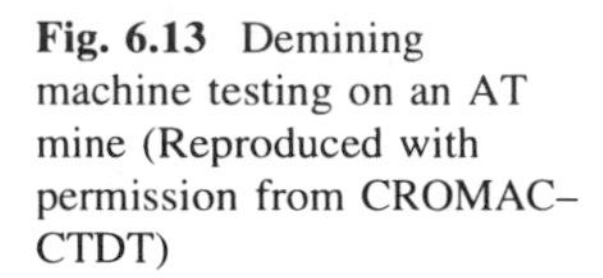
Fig. 6.13 Demining machine testing on an AT mine (Reproduced with permission from CROMAC–CTDT)

controlled. If machine is equipped and steered from the cockpit, machine operator protection against noise and acceleration variation when AT mine TMA-3 is activated, is performed. Within cockpit, instruments for noise measurement and acceleration measurement are mounted. Laid TMA-3 AT mine is, from shelter and safe distance, activated electrically. If impulse noise within cockpit is higher than allowed (120 dB, pain level is 100 Pa), improvement of cockpit isolation or wearing of ear protectors, can be requested to be worn during machine operation. Highest value of acceleration variation and its influence on operator's legs should be less than 3 m/s.

6.5.6 Demining Machine Usage Acceptability

Acceptability of demining machines is determined after the performance and machine survivability testing has been conducted.

An example of rejection and acceptance

On top of laid fibreboards, the cutting depth of a working tool's is measured across the whole length of the tool. A mean value of all three fibreboards in the test lane is taken for the assessment of soil digging depth. If there are one or more ''spaces'' (areas not affected by cutting) existing on the fibreboards of a length greater than the diameter of at least that of a AP mine, the soil digging depth profile, i.e. soil digging density is not satisfactory.

If the mean value of the measured soil digging depth in test lanes of the I. category of soil is greater than 17 cm and smaller than 19.5 cm, with the condition that none of the single measurements is less than 12 cm, the demining machine can obtain the grade that it successfully processes soil in the I. category of up to 20 cm.

6.5.6.1 Unreliability of Mine Neutralization

For machine acceptability, all mines set according to the standard [2] have to be neutralized. If the mine remained undamaged, the procedure needs to be repeated with that specific mine type (in the complete scheduled number). If after a repeated demining machine acceptability testing procedure with AP and/or AT mines again one or more mines remain unneutralized, the working tool does not meet the set standards.

6.5.6.2 Assessment of Acceptability

On basis of a machine testing report a conclusion is derived. For each testing lane the results are given: average digging depth, machine passing time, machine movement speed and the estimated machine efficiency. The reliability of AP and AT mine neutralization is also given, machine endurance as well as machine operator protection.

New machine designs in addition can go through operative testing for specific local soil categories in request of a customer (cca. 3 ha—light machines, 5 ha—medium machines, 8 ha—heavy machines), within a realistic project of humanitarian demining, in order to affirm the performance and efficiency results. In the end, an acceptability or unacceptability assessment of the demining machine for the mechanic treatment of mine suspected area in specific conditions of humanitarian demining is decided.

6.5.7 Annual Compliance Testing

The goal of annual machine compliance testing is:

- to test technical machine characteristics
- to test mine clearing performances.

Machines used in demining are being worn out, which requires replacement of certain machine or working tool parts that have important role in machine clearance quality, which further influence total machine operation results. If machine or working tool have been changed in design or quality, machine is fully tested again. Annual compliance/compatibility testing of light and medium demining machines and excavators is performed on field machine test range, and heavy machines on the demining project area, according to annual demining operation plan.

Annual machine compliance testing comprises:

- testing of technical machine characteristics,
- machine testing on test lane containing local soil–soil digging depth profile of a flail or tiller machine.

6.5.8 Conclusion

Demining machine testing is constant process that should follow up the development of SOP, as well as development and use of new machines. The following results were obtained during demining machine testing:

- Machine users are provided with tested and high quality demining machines,
- Systematic database is available for use to users and monitoring service,
- Possibility of use of those machines that provide optimal results on certain demining project,
- In Croatia, in 90 % of demining projects, demining machines were used,
- Demining process is significantly speeded up,

6.6 Research of Machine Usage on Soft Soil

Demining machines treat mine suspected soil mechanically and, in so doing, neutralize buried mines (by crushing them or activating). Demining machine usage depends on condition of soil that needs to be demined. Machine mobility evaluation, from the viewpoint of their demining usage possibilities, is based on soil trafficability analysis.

Soil trafficability is the criterion of possible machine usage in demining. The machine that treats the soil with a flail or tiller digs deep to 20 cm or less, and by that forms a new loose layer of certain thickness in front of the tracks or wheels, through which the machine further drives, Fig. 6.14.

Therefore, it is necessary for the machine to have the required technical parameters of tracks and wheels for work in such hard conditions, particularly on that new moist layer of soil. The machine needs to keep up its working speed and controllability, without slipping and course violation. The new loose layer of soil ("banana layer") on which the track or wheel proceeds, has very low soil capacity, thereby a problem occurs with trafficability of such a soil layer.

In practice, in accordance with SOP CROMAC [4], demining machines are not used in the following cases:

- When the soil is frozen and/or covered in snow,
- In case of rain or thick fog,
- When operation of demining machine is limited for the reasons of inadequate soil and climate conditions,
- When it is not possible to ensure machine operating from a supporting vehicle at the safe distance, or from the machine cockpit

Problems:

- Machine usage on moist coherent soil is not sufficiently determined
- Unsafe hold of machine trace on loose layer of soil (depth, direction, turn)

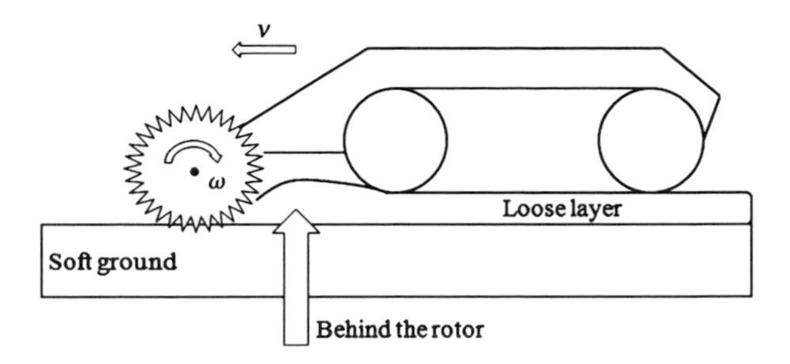

Fig. 6.14 Forming of loose layer of soil under the tracks

- Mines being pressed deeper in the soft soil
- Machine usage is dependent on implementer's experience (demining company)

6.6.1 New Loose Soil Layer—"Banana Layer"

In the domain of terramechanics and machine mobility on coherent soil, soil trafficability of very moist loose soil hasn't yet been explored. It is very important that the methodology of machine usage research in these conditions is set. Machine usage limitations can come from the side of soil moisture, which utmostly influences the machine usage possibility, and also the impossibility of activating mines on soft soil. This problem exists also for the supporting vehicle, which is usually on wheels. The task is to explore this field of machine use possibility and set out the models of anticipating ''banana layer'' trafficability under the machine [5].

Goals of machine usage research on soft soil

1. Examine the problem of loose soil layer trafficability at which the machine moves (soil load capacity, machine pressure on soil, mover sinkage, adhesion)
2. Determine the machine usage acceptability model in humanitarian demining
3. Define the machine usage evaluation procedure

Assumption

Machine usage criterion is its mobility on soft ground. Mobility of the machine, encompasses:

1. Soil trafficability
2. Working machine soil treatment speed (flat ground, climb, inclination)
3. Keeping the trace (line, digging depth, manoeuvre)

Soil conditions

Conditions defined by soil type and category, terrain configuration and vegetation are referred to as soil conditions. In relation to probing and demining there are:

- Favourable soil conditions,
- Aggravated soil conditions,
- Difficult soil conditions,
- Specific conditions.

Moisture content
The moisture content of a soil is expressed as a percentage of the dry mass:
Moisture content, w = loss of moisture/dry mass × 100 %
w = (wet mass–dry mass)/dry mass × 100 %

6.6.2 Soil Trafficability in the Domain of Terramechanics

In theory of evaluating machine movement, a couple of approaches are used in researching soil trafficability:

- on the basis of nominal pressure on soil—NGP,
- on the basis of calculating mean maximum pressure on soil—MMP,
- on the basis of calculating the cone index of soil—CI and CI_L,
- on the basis of cone index of the vehicle—VCI,
- on the basis of *Bekker* theory of soil constraint, and
- on the basis of wheel and soil slippage analysis.

Nominal pressure on soil—NGP is the marginal tangential pressure of wheel on soil, which doesn't provide us with relevant comparison between two different wheels because of neglecting the tyre deformation influence of the laden wheel whilst in movement. Similarly, it can be applied to tracks (different shapes and parameters of tracks). Relevant indicators of terrain vehicle mobility assessment, both on tracks and wheels, are based on the British model of MMP pressure and the American model of VCI pressure which was adopted as the referent NATO model of evaluating military machine mobility, on wheels and tracks—NRMM–*NATO Reference Mobility.* However, practically there still doesn't exist an explained and defined machine movement on loose soil of high moisture.

Demining machine's mass complying to trafficability requirements

Machine mass influence is very significant in trafficability indicators of MMP and VCI. Machine mass needs to be adjusted to soil load capacity. Cone index for very soft soil is $CI < 300$ kPa. The load capacity of such soil depends on soil moisture (w %). For soft soils, the load capacity ranges from 50 (100) kPa to 300 kPa. That means all machines (light, medium or heavy) can have mean maximum pressure on soil to, at most, 300 kPa. According to the above mentioned, the greatest allowable demining machine mass, on soft terrain, can be determined. Depending on the cone index, machines can be quickly evaluated. However, there always exists the problem of measuring low soil capacity in high moisture conditions.

Soil trafficability analysis, because of machine usage evaluation, is conducted on the basis of comparison between soil load capacity and mean maximum machine pressure on soil.

Interaction between soil and wheels and tracks, includes:

1. Soil consistency
2. Wheels and tracks
3. Machine pressure on soil
4. Deformation of soil under the wheel
5. Soil trafficability simulation
6. Model of machine usage acceptability

6.6.3 Conclusion

Digging the soil using flail or tiller in front of the wheels or tracks creates a new loose layer of soil, and a machine has to move over it. Due to the higher level of soil moisture, mobility of the machine is limited, and its work efficiency decreases. Mobility of the machine depends on soil trafficability and working speed of movement of the machine. In the Theory of the machine movement, the soil trafficability has not been researched and explained, yet. Hence, the use of the machine on dry, wet, and extremely wet soil has not been sufficiently defined, so the use of the machine strictly depends on the experience of the evaluating person. It often occurs that the machine, in real mine field situation, does not hold the lane well, that it does not have satisfactory working speed of movement, that it does not have sufficient controllability, so it sways, or skids off the tracks or even rolls over. Occasionally in such conditions, the mine is pressed even deeper into soil. Such sort of machine demining is of poor quality, and thus it is unacceptable. The final goal is to evaluate the use of demining machines in relation to actual state of soil.

Soil trafficability needs to be estimated on demining projects, on demining machine usage guidelines in various conditions and seasons. Therefore, soil trafficability has to be distinguished and evaluated in all categories of machines and supporting vehicles. Models for calculating machine mobility need to be compiled. Maps of machine movability on certain terrain have to be created, in accordance with their usage assessment in mine-suspected area.

References

1. Test and evaluation of demining machines (2009) CEN Workshop Agreement, CWA 15044:2009, Supersedes CWA 15044:2004, CEN, Brussels.
2. Humanitarian demining—Requirements for machines and conformity assessment for machines (2009) Standard HRN 1142, Croatian Standards Institute, HZN 1/2010, Zagreb.
3. Axelsson H, Sundqvist O (2003) Mine Clearance Vehicles, Crew Safety Standard, FMV, The Swedish Defence Material Administration, Test Range Karlsborg, Stockholm.
4. Ordinance on Technical Requirements and Conformity Assessment of Devices and Equipment Used for Humanitarian Demining (2007), National Gazette no. 53/07.
5. Mikulic D, Koroman V, Ambrus D, Majetic V (2007) Concept of Light Autonomous Machines for Dual Use, Proceedings of the Joint North America, Asia-Pacific ISTVS Conference and Annual Meeting of Japanese Society for Terramechanics, University of Alaska Fairbanks, Alaska, Fairbanks.

Abbreviations

AP	Anti-personnel mine
AT	Anti-tank mine
CEN	European Committee for standardization
CG	Center of gravity
CROMAC	Croatain Mine Action Centre
CROMAC CTDT	Croatain Mine Action Centre—Centre for Testing, Development and Training
CWA	CEN Workshop Agreement
GICHD	Geneva International Centre for Humanitarian Demining
IMAS	International mine action standards
LAM	Light autonomous machines
LDM	Light demining machine
MAC	Mine action centre
MSA	Mine suspected area
OCU	Operator control unit
OHU	Operator hand-held unit
PROM-1	Bouncing AP mine
QA	Quality assurance
QC	Quality control
SOP	Standard operative procedures
TDM	*Tele-demining* machine
UXO	Unexploded ordnance

D. Mikulic, *Design of Demining Machines*,
DOI: 10.1007/978-1-4471-4504-2, © Springer-Verlag London 2013

Index

D. Mikulic, *Design of Demining Machines*,
DOI: 10.1007/978-1-4471-4504-2, © Springer-Verlag London 2013

V

W

Printed by Publishers' Graphics LLC
MO20121107.19.23.26